Starry Night

Starry Night

Astronomers and Poets Read the *Sky*

David H. Levy

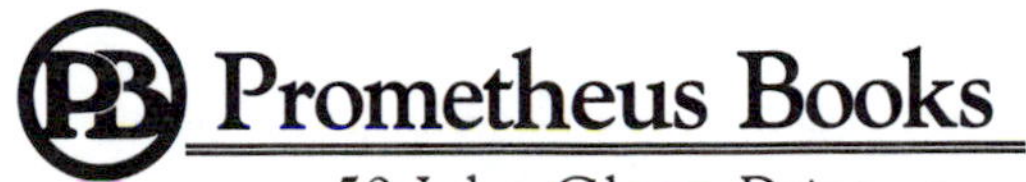

59 John Glenn Drive
Amherst, New York 14228-2197

This book is a completely revised edition of *More Things in Heaven and Earth: Poets and Astronomers Read the Night Sky*, published by Wombat Press.

The publisher greatly acknowledges the cover art composition by Steven Slipp.
Cover images: Aurora Borealis (photo by Leo Enright), Comet Shoemaker-Levy 9 (photo by David Levy), and the Moon (photo by David Levy)

Published 2001 by Prometheus Books

Inquiries should be addressed to
Prometheus Books
59 John Glenn Drive
Amherst, New York 14228-2197
VOICE: 716-691-0133, ext. 207
FAX: 716-564-2711
WWW.PROMETHEUSBOOKS.COM

05 04 03 02 01 5 4 3 2 1

Library of Congress Cataloging-in-Publication Data

Levy, David H., 1948–
[More things in heaven and earth]
Starry night : astronomers and poets read the sky / David H. Levy.
p. cm.
Previously published in 1997 under the title: More things in heaven and earth.
Includes bibliographical references and index.
ISBN 1-57392-887-9 (pbk. : alk. paper)
1. English poetry—History and criticism. 2. Astronomy in literature.
3. American poetry—History and criticism. 4. Literature and science—History. I. Title.

PR508.A76 L48 2001
821.009'356—dc21 00-045916

Printed in the United States of America on acid-free paper

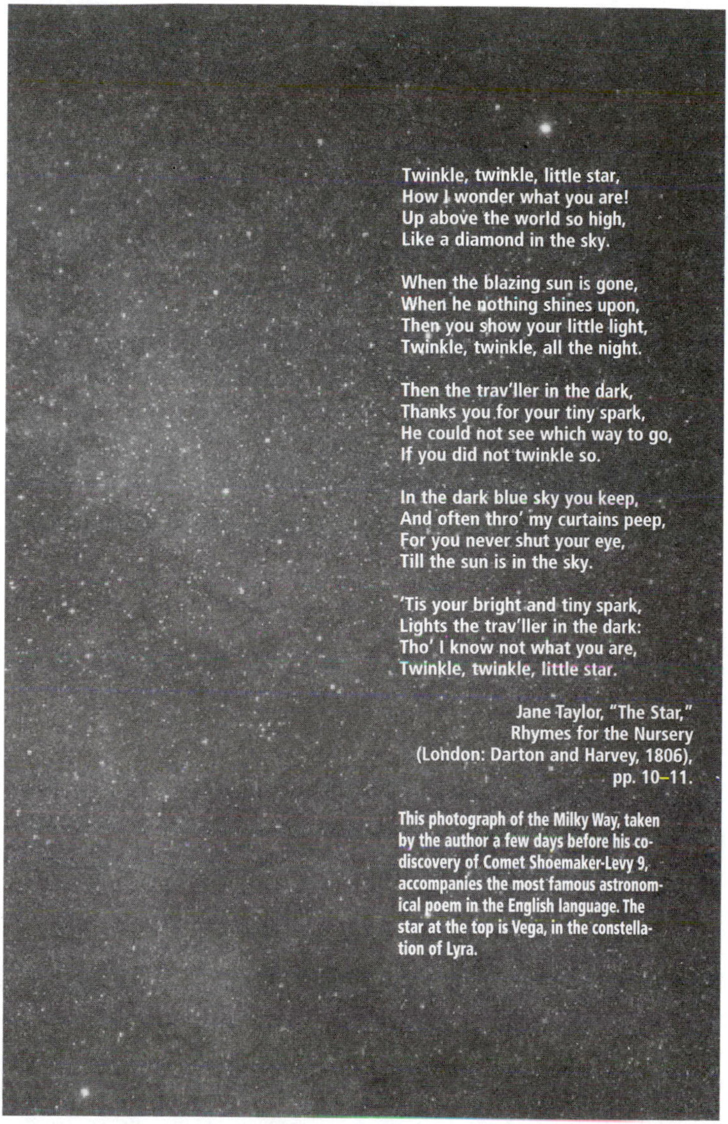

Twinkle, twinkle, little star,
How I wonder what you are!
Up above the world so high,
Like a diamond in the sky.

When the blazing sun is gone,
When he nothing shines upon,
Then you show your little light,
Twinkle, twinkle, all the night.

Then the trav'ller in the dark,
Thanks you for your tiny spark,
He could not see which way to go,
If you did not twinkle so.

In the dark blue sky you keep,
And often thro' my curtains peep,
For you never shut your eye,
Till the sun is in the sky.

'Tis your bright and tiny spark,
Lights the trav'ller in the dark:
Tho' I know not what you are,
Twinkle, twinkle, little star.

Jane Taylor, "The Star,"
Rhymes for the Nursery
(London: Darton and Harvey, 1806),
pp. 10–11.

This photograph of the Milky Way, taken by the author a few days before his co-discovery of Comet Shoemaker-Levy 9, accompanies the most famous astronomical poem in the English language. The star at the top is Vega, in the constellation of Lyra.

For Wendee

Who gave me the greatest gift,

The opportunity to share our lives.

Contents

What is it all but a trouble of ants in the gleam of a million million of suns?

—Alfred, Lord Tennyson, "Vastness" (1885), The Poems of Tennyson, ed. Christopher Ricks (New York: W. W. Norton, 1969), p. 1346.

The Whirlpool Galaxy is an example of a great spiral galaxy viewed face-on from Earth. Photo by Tim Hunter.

Acknowledgments

The *seed for* this book was planted in 1970, as I day-dreamed in Roger Lewis's Victorian Poetry class at Acadia University. We were reading *In Memoriam*, and Roger pointed out Tennyson's incredible grasp of the science of his time. Throughout these intervening years, I have always appreciated that moment when I first became aware of the interplay between science and poetry. A quarter century later, Roger and I discussed the idea of a nonscholarly book aimed at showing this relationship. Had it not been for our continuing friendship, I probably would never have thought of writing this book.

In 1978, Norman MacKenzie, of Queen's University, Kingston, introduced me to Gerard Manley Hopkins's poetic fragment "I am like a slip of comet." More than any other piece of literature I have read, I consider these lines to contain the real substance of the relation of astronomy to literature. I thank Norman again for his careful suggestions on chapters 4 and 5 of this book, and for his especially helpful suggestions for this new edition. I am also grateful for Rita McKenzie's kind permission to use her original painting of Gerard Manley Hopkins in chapter 5. She spent many hours on this original interpretation of one of my favorite English poets.

I want to thank Steven Slipp for his invaluable suggestions regarding the content, design, and production, and to Margo Grant for her page design and layout. Jan Williams's copyediting was superb; it sharpened the prose and the flow. Additionally, I thank Joan-ellen Rice for providing useful information; Mary Lou and David Henry for commenting on the first chapter; and Acadia President Kelvin Ogilvie for contributing the foreword. For this new edition I thank Linda Regan, Art Merchant, and Prometheus Books for their excellent suggestions for improvements. I am also grateful for the prompt assistance of Ellen Meltzer, of the University of California of Berkeley's Moffitt Library, who found a rare copy of Sarah Williams's *Twilight Hours: A Legacy of Verse,* with its copy of "The Old Astronomer." Finally, a special hug for my wife, Wendee, whose help with copyediting was invaluable, and whose love and encouragement throughout the writing of this book enabled me to bring it to completion.

Foreword

T*he heavens have* captivated the curiosity, the imagination, and the fears of life on earth throughout the eons of evolution. One can imagine the first form of life with the powers of observation being perplexed by magnificent cosmic displays. Think of the great herds of dinosaurs arching their heads skyward, perhaps pondering some magnificent change in the heavenly spectrum. Humanoids from their very beginning have marveled, feared, interpreted, and used to advantage the wondrous changes of the sky. To gods of all sorts have been attributed the powers presumed to be lying in the firmament. Throughout it all the senses of humankind inspired and articulated the characteristics of curiosity, emotion, study, analysis, creativity—the continuum that evolved the arts and the sciences.

The magnetic pull of the heavens with their apparent infinite depth, the gentle twinkling of stars, the wonderful displays of "shooting stars," the appearance of the great comets,

A break in the clouds reveals a comet. Photo by David Levy of the comet he discovered in 1990.

All this long eve, so balmy and serene,
Have I been gazing on the western sky,
    And its peculiar tint of yellow green:
And still I gaze—and with how blank an eye!
And those thin clouds above, in flakes and bars,
That give away their motion to the stars;
Those stars, that glide behind them or between,
Now sparkling, now bedimmed, but always seen:
Yon crescent Moon, as fixed as if it grew
In its own cloudless, starless lake of blue;
I see them all so excellently fair,
I see, not feel, how beautiful they are!

—Samuel Taylor Coleridge, "Dejection: An Ode" (1802); English Romantic Poetry and Prose, ed. Russell Noyes (New York: Oxford University Press, 1956), p. 396.

the Milky Way, the constellations and the planets, and the sound and the fury of electrical storms—all these things have played on the fears of the human race. Why are we here? Where did we come from? Give us a sign. The heavens have far too often complied with signs that have pointed to the best and the worst of human nature.

We have sought to explain these things. Artists have written, painted, and brought through poetic expression all of the elements of the study of this vast unknown. Some of these artists have been called scientists as they pursued the stepwise analysis of the observations of the universal ether. The scientific method has added breadth to the great artistic pursuit of the unknown. Together, all of the curious have brought to us the detailed charts of the heavens, the travels of earth launched space craft, the awareness of black holes and red dwarfs, and still we are fascinated by the expanding mystery of the universe.

David Levy has pursued his dream. An earthbound explorer lifted into the heavens by the power of the telescope and by his own curiosity, David needed another critical human characteristic—the self confidence to make the mortal decision to gaze heavenly with the determination to look, to observe steadily, and to record. Can anyone doubt the character required of a young man standing at his telescope, gazing at the heavens for an entire night through the clear skies offered on the Acadia campus? What must his fellow students have thought?

In the mold of scholars, artist-scientist Levy is now inscribed in those very heavens, traveling the orbit of twenty-one comets. With nature as artist, Maestro Levy commanded a worldwide orchestra as the most spectacular heavenly display ever recorded unfolded when Shoemaker-Levy 9 blended its more than twenty legions into the atmosphere of Jupiter with cosmic grace and fury. The pens of poets, philosophers, scribes, and scientists move together to express the capacity of the human spirit lifted to the cosmos.

David Levy is a natural symbol of the oldest of truths, the continuum of artistic and scientific expression. From David's story let our universities learn again this essential truth. By lifting our spirits to such majestic heights perhaps he can help put back together on earth what mere mortals have torn asunder, the union of art and science.

Kelvin K. Ogilvie, C.M. Ph.D., D.Sc., F.C.I.C.
President of Acadia University

Preface

Come, Gentle Night

T*wo men, one* a writer, the other a geologist, now reside next to each other on the Moon. The geologist is Gene Shoemaker. In the spring of 1948, as a young Caltech graduate working in a small Colorado town for the U.S. Geological Survey, Shoemaker had a dream. Within my lifetime, he thought, humanity will be landing on the Moon. Now what kind of person should be on the Moon? A geologist, of course. On that day Shoemaker's compass was set, his journey to the Moon started. In the next fifteen years Shoemaker distinguished himself in the study of impact craters, both here on Earth and on the Moon. With some three hundred such craters visible through a small telescope all across the Moon, Shoemaker held that the dominant geological process there was impacts from comets, asteroids, and small pieces of dust called meteoroids.

In 1963, while he was standing at the head of the line to be chosen to lead a team to the Moon, Shoemaker was struck with Addison's disease, a condition that shuts down the adrenal cortex. Although by the summer of 1963, the condition was under good medical control, Shoemaker knew that he would never pass the medical exam required of astronauts. "No," Shoemaker thought, "I'm not going to make it to the Moon. When you had that idea in your head for fifteen years, it doesn't go away. I couldn't help feeling that there, but for that failed adrenal gland, go I. For a long time after, I used to have dreams—thought I got there—got to the Moon, was there doing geology. I had to go do other things."[1]

Of the many "other things" that Gene Shoemaker set his energy to, the most important was a study not of the craters, the results of impacts carved in stone, but of the comets and asteroids that have the potential to make those craters. I had the singular honor of joining both Shoemakers, Gene and Carolyn, in their search project. We found thirteen comets together. One of those comets, Shoemaker-Levy 9, actually demonstrated the role of collisions in our solar system by smashing into Jupiter. The result was the greatest explosion ever seen in the solar system by humanity. As Leslie Peltier, an Ohio farmboy who discovered a dozen comets, wrote years earlier, comets can teach us about our world in a special way:

> Time has not lessened the age-old allure of the comets. In some ways their mystery has only deepened with the years. At each return a comet brings with it the questions which were asked when it was here before, and as it rounds the sun

> and backs away toward the long, slow night of its aphelion, it leaves behind with us those questions, still unanswered.
>
> To hunt a speck of moving haze may seem a strange pursuit, but even though we fail the search is still rewarding, for in no better way can we come face to face, night after night, with such a wealth of riches as old Croesus never dreamed of.[2]

The Shoemakers were not content just to search for comets and asteroids; they also took annual field trips to the Australian outback to study the rich variety of impact sites located there. During one of these trips, on July 18, 1997, their vehicle was struck head-on. Although Carolyn survived the accident, Gene was killed instantly.

Gene Shoemaker spent his life studying the confluence of events that would lead to the joining of worlds, like comets and planets, or asteroids and planets. Through his death, a confluence was taking place that would involve not physical worlds but spheres of study. In July 1997, final preparations were underway to launch a spacecraft called Lunar Prospector. After a three-day trip to the Moon, the spacecraft would orbit it for about a year, then crash into the lunar surface. As the scientific world recoiled in shock over Gene Shoemaker's death, well-known planetary scientist Carolyn Porco had the idea to let Gene Shoemaker achieve in death what eluded him in life.

As the first quarter Moon dominated the night sky on the

evening of January 6, 1998, an Athena 2 rocket ignited in a burst of flame and soared into the night. Tucked inside the spacecraft was Porco's idea, a tiny canister containing a small sample of the ashes of Gene Shoemaker. Surrounding them was a tightly wound thin foil on which was inscribed part of this passage from Shakespeare's *Romeo and Juliet*:

> *Come, gentle night, come, loving, black-brow'd night,*
> *Give me my Romeo, and, when he shall die,*
> *Take him and cut him out in little stars,*
> *And he will make the face of heaven so fine*
> *That all the world will be in love with night,*
> *And pay no worship to the garish sun.*[3]

Since October 4, 1957, when the Soviet Union launched their first Sputnik, there have been hundreds of spaceflights. Two of these flights, Voyager 1 and 2, even carry collections of writing from the people of planet Earth as they speed out to the stars. When Lunar Prospector crashed onto the Moon in July 1999, its successful mapping mission completed, it became the first craft to carry the words of William Shakespeare, probably the most significant writer to put pen to paper, to another world.

Beyond the Equations

William Shakespeare deserves a place among the stars. The astronomical references in his writing, for one reason, show a considerable knowledge both of the night sky and his understanding, as expressed through his characters, of the relation between humanity and nature. "There are more things in Heaven and Earth," said Hamlet, "than are dreamt of in your philosophy." Since I was brought up by parents who admired Shakespeare, Hamlet's wise and famous observation about nature became an integral part of my youth. We never saw the world as a battleground between the opposing forces of arts and science. We saw it as a harmonious place, with lots of room for varied interests. When at age twelve, my interest soared to the stars, my family encouraged my love of reading and literature to go along for the ride.

Throughout my early years of stargazing, I always enjoyed the times when a look through a telescope would remind me of some famous poem. One early morning I arose before dawn to draw a chart of the stars in the Pleiades. Through a telescope these stars were not seven, but at least ten times seven, a perfect match for Tennyson's description of them glittering "like a swarm of fire-flies tangled in a silver braid."

Despite the poetry that was forever looking my way though the other end of my telescope, by the end of my high school days, when I wanted to be a professional astronomer, I felt a pull away from literature. As a physics major at McGill University in Montreal, I was impressed by the single-minded sense of the scientist—like Marie Curie huddled in her lab working on the discovery of radium—we scientists had no time for literary distraction. And as beautiful as a poem might be, it is a distraction. Tennyson and Shakespeare books were closed.

In 1968, chastened by an inability to do well in physics and chemistry, I transferred to Acadia University, in the rural town of Wolfville, Nova Scotia. There was nothing esoteric about nature at Acadia; the University stood near the tortured shore of the Minas Basin, whose fifty-foot tides are the most dramatic on Earth, and whose night sky, far from any large city, is pristine. Although my career direction was still in science, the English literature courses I took were tantalizing.

Is there a way to appreciate science beyond the equations? I found the answer to that one while taking second-year English, a survey course with Roger Lewis. During his course my question was answered as clearly as the clang of the wild bells in Tennyson's famous elegy *In Memoriam*. In some 130 sections, this poem offered a completely new way of looking at the evolution of cosmos and humanity. *In Memoriam* was science without the footnotes. In Tennyson's world, the two Charles's—Sir Charles Lyell and Charles Darwin—were the state of the art. Lyell's geology showed an Earth whose history was not marked by violent change but by the steady state of slow, orderly development. Darwin applied the teachings of geologic uniformitarianism to the origin and evolution of life. Many scientific papers were written during the nineteenth century to support, or not to support, these theories, but the one document that has survived the time to land in the college classes of today is not a science paper at all but that masterful, exquisite poem called *In Memoriam*. That second-year course was my epiphany; I could now transfer freely from a geology major to an English major without losing my anchor in science. In 1972 I graduated with a major in English, a minor in Geology, and a passion for the night sky.

Comet Shoemaker-Levy 9, taken by Wieslaw Wiesniewski on March 28, 1993, just three days after discovery.

A few years later, I was starting a master's degree at Queen's University in Ontario, Canada. Under Norman MacKenzie I discovered the astronomical magic of Gerard

Manley Hopkins, a nineteenth-century English poet whose youth had been punctuated by a train of bright comets. Dr. McKenzie told me of a poem that Hopkins wrote about the passage of a comet. The poet had intended it to be part of a play, a soliloquy in which the character compared her life to the passage of a great comet traveling from the depths of the outer solar system, sighting the Sun and glowing with its energy, then finally turning about and fading away. Dr. MacKenzie encouraged my work on this theme of poetry and the night sky.

I was so taken with Hopkins's brilliant merging of careful observation of the night sky and his insightful poetry that I decided to concentrate on that poem, and its author, during my graduate years. After I completed my work at Queen's, I moved to the clear sky and desert environment near Tucson, Arizona, where over the next two decades my discoveries of twenty-one comets cemented my feelings and thoughts about the relation between literature and science.

The book you are about to read focuses on poetry and the night sky, an important subset of literature and science. To thoroughly understand this relation, it helps to have an appreciation for both fields. To appreciate the night sky, it helps to go outdoors and, as Hopkins wrote, "look at the stars." It is my hope that through this book, you will find or make your own connection between the words of writers like William Shakespeare and those distant points of light we see when we look to the sky. After all, thanks to a merging of spacecraft technology and the imagination, William Shakespeare and Gene Shoemaker have found a way to merge the arts and sciences on a distant world.

The Old Astronomer

Reach me down my Tycho Brahe,—I would know him when we meet,
When I share my later science sitting humbly at his feet;
He may know the law of all things, yet be ignorant of how
We are working to completion, working on from then till now.

Pray, remember, that I leave you all my theory complete,
Lacking only certain data, for your adding, as is meet,
And remember, men will scorn it, 'tis original and true,
And the obloquy of newness may fall bitterly on you.

But, my pupil, as my pupil you have learnt the worth of scorn,
You have laughed with me at pity, we have joyed to be forlorn;
What, for us, are all distractions of men's fellowship and smiles?
What, for us, the goddess Pleasure, with her meretricious wiles?

You may tell that German college that their honour comes too late.
But they must not waste repentance on the grizzly savant's fate;

Though my soul may set in darkness, it will rise in perfect light;
I have loved the stars too truly to be fearful of the night.

What, my boy, you are not weeping? You should save your eyes
for sight;
You will need them, mine observer, yet for many another night.
I leave none but you, my pupil, unto whom my plans are known,
You "have none but me," you murmur, and I "leave you quite
alone"?

Well then, kiss me,—since my mother left my blessing on my
brow,
There has been a something wanting in my nature until now;
I can dimly comprehend it,—that I might have been more kind,
Might have cherished you more wisely, as the one I leave
behind.

I "have never failed in kindness"? No, we lived too high for
strife,—
Calmest coldness was the error which has crept into our life;
But your spirit is untainted, I can dedicate you still
To the service of our science—you will further it? You will!

There are certain calculations I would like to make with you,
To be sure that your deductions will be logical and true;
And remember, "Patience, Patience," is the watchword of a sage,
Not to-day nor yet to-morrow can complete a perfect age.

I have sown, like Tycho Brahe, that a greater man may reap;
But if none should do my reaping, 'twill disturb me in my sleep.
So be careful and be faithful, though, like me, you leave no name;
See, my boy, that nothing turn you to the mere pursuit of fame.

I must say Good-bye, my pupil, for I cannot longer speak;
Draw the curtain back for Venus, ere my vision grow too weak:
It is strange the pearly planet should look red as fiery Mars,—
God will mercifully guide me on my way amongst the stars.

—Sarah (Sadie) Williams, 1869[4]

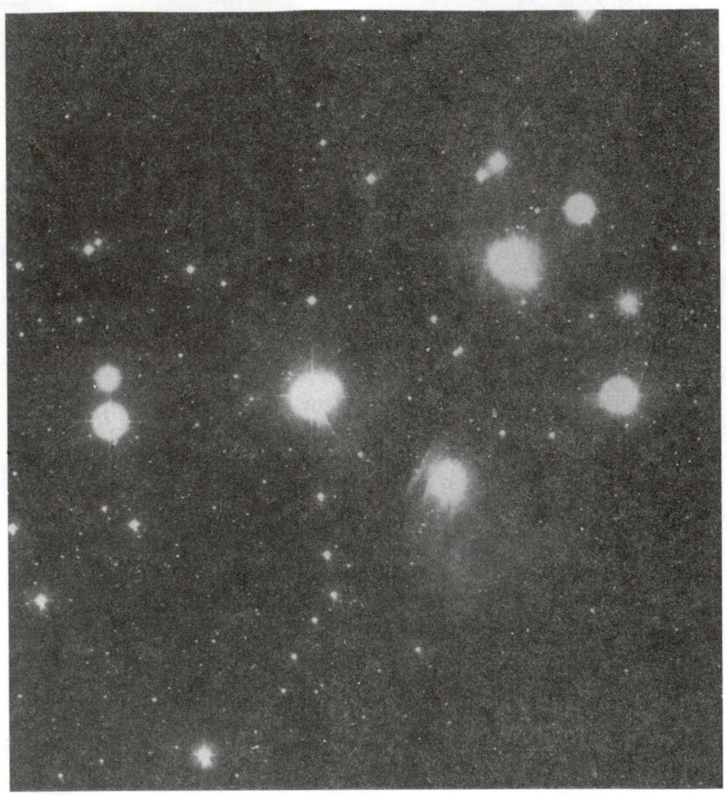

How noteless Men, and Pleiads, stand,
Until a sudden sky
Reveals the fact that One is rapt
Forever from the Eye—

Members of the Invisible,
Existing, while we stare,
In Leagueless Opportunity,
O'ertakeless, as the Air—
Why didn't we detain Them?

The Heavens with a smile,
Sweep by our disappointed Heads
Without a syllable—

—Emily Dickinson, Poem 282 (1861),
The Complete Poems of Emily Dickinson,
ed. Thomas H. Johnson
(Boston: Little, Brown, 1890, 1960), p. 130.

The Pleiades.
Photo by Eugene and Carolyn Shoemaker and David Levy.

On a starred night Prince Lucifer uprose.
Tired of his dark dominion swung the fiend . . .
He reached a middle height, and at the stars,
Which are the brain of heaven, he looked, and sank.
Around the ancient track marched, rank on rank,
The army of unalterable law.

—George Meredith, "Lucifer in Starlight" (1883),
Victorian Poetry and Poetics,
ed. Walter E. Houghton and
G. Robert Stange
(Boston: Houghton Mifflin Co., 1968), p. 651.

Two men look out through the same bars:
One sees the mud, and one the stars.

—Frederick Langbridge,
"A Cluster of Quiet Thoughts"
(1849–1923),
(Religious Tract Society Publication),
Oxford Dictionary of Quotations
(London: Oxford University Press, 1955),
p. 310.

The Pillars of Creation, showing a new star, with solar system, in formation. NASA/Hubble Space Telescope photograph.

Follow wise Orion
Till you waste your Eye—
Dazzlingly decamping
He is just as high—
—Emily Dickinson, Poem 1538 (1882), The Complete Poems of Emily Dickinson, ed. Thomas H. Johnson (Boston: Little, Brown, 1890, 1960), p. 130

The Great Nebula in Orion. Photo by Rik Hill of Tucson, Arizona.

Read Nature; Nature is a friend to truth . . .
Hast thou ne'er seen the comet's flaming flight?
The' illustrious stranger, passing, terror sheds
On gazing nations, from his fiery train
Of length enormous; takes his ample round
Through depths of ether; coasts unnumber'd worlds
Of more than solar glory; doubles wide
Heaven's mighty cape; and then revists earth,
From the long travel of a thousand years.
Thus, at the destined period, shall return . . .
—Edward Young (1683–1765), "Night Thoughts," Selected Poems, ed. B. Hepworth (Cheadle, England: Fyfield Books, 1975), pp. 93–94.

Comet Hyakutake and a saguaro cactus plant. Photo by David Levy.

This Hubble Space Telescope Deep Space Image shows galaxies farther away than anything shown before. NASA image.

When the Astronomer stops seeking
For his Pleiad's Face—
When the lone British Lady
Forsakes the Arctic Race

When to his Covenant Needle
The Sailor doubting turns—
It will be amply early
To ask what treason means.

—Emily Dickinson, Poem 851 (1864),
The Complete Poems of Emily Dickinson,
ed. Thomas H. Johnson (Boston: Little, Brown, 1890, 1960), p. 130

Till the sun grows cold,
And the stars are old,
And the leaves of the Judgement Book unfold!
—Bayard Taylor, "Bedouin Song: Refrain,"
The Poems (Boston: Ticknor and Fields, 1866),
pp. 134–35.

Sky from Jarnac Pond.
Photo by David Levy.

Opposite: The eastward-moving earth makes the stars circle the pole in this time exposure photo by Rik Hill.

Move eastward, happy earth, and leave
Yon orange sunset waning slow;
From fringes of the faded eve,
O happy planet, eastward go,
Till over thy dark shoulder go
Thy silver sister-world, and rise
To glass herself in dewy eyes
That watch me from the glen below.
Ah, bear me with thee, smoothly borne,
Dip forward under starry light,
And move me to my marriage-morn,
And round again to happy night.
—Alfred, Lord Tennyson
"Move Eastward, Happy Earth" (1842),
Victorian Poetry and Poetics,
ed. Walter Houghton and G. Robert Stange
(Boston: Houghton Mifflin Co., 1968), p. 42.

I am the owner of the sphere,
Of the seven stars and the solar year,
Of Caesar's hand, and Plato's brain,
Of Lord Christ's heart, and Shakespeare's strain.
—Ralph Waldo Emerson (1803–1882),
"The Informing Spirit,"
The Poems of Ralph Waldo Emerson
(New York: The Heritage Press, 1945), p. 183.

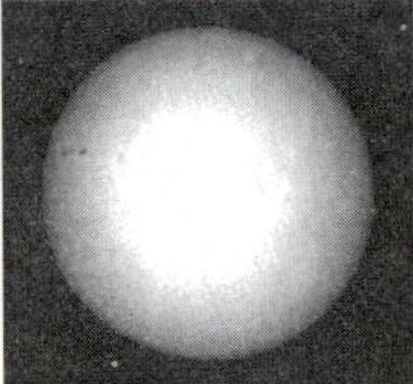

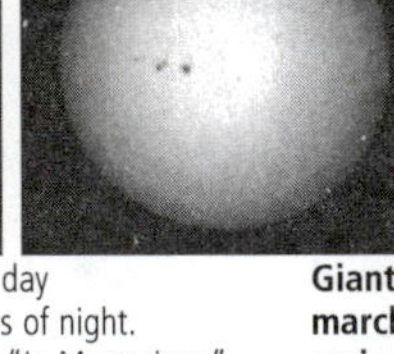

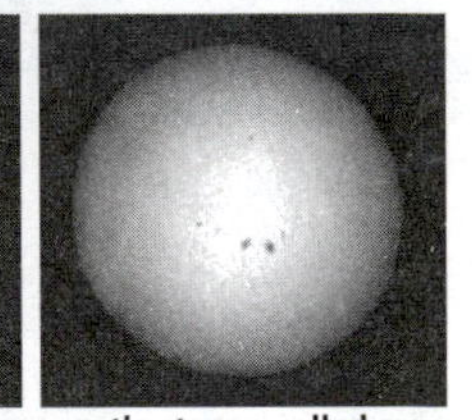

The very source and fount of day
Is dash'd with wandering isles of night.
—Alfred, Lord Tennyson, "In Memoriam,"
Victorian Poetry and Poetics,
ed. Walter Houghton and G. Robert Stange
(Boston: Houghton Mifflin Co., 1968), p. 76.

Giant magnetic storms called sunspots march across the face of the Sun in this series of photographs by Leo Enright of Sharbot Lake, near Kingston, Ontario.

The Aurora Borealis. Photo by Leo Enright.

As lines, so loves oblique, may well
Themselves in every angle greet:
But ours, so truly parallel,
Though infinite, can never meet.

Therefore the love which us doth bind,
But Fate so enviously debars,
Is the conjunction of the mind,
And opposition of the stars.

—Andrew Marvell (1621–1678), "The Definition of Love," Renaissance England: Poetry and Prose from the Reformation to the Restoration (New York: W. W. Norton, 1956), p. 997.

Does this ageless sonnet evoke the vastness of this great and distant spiral galaxy? NGC 891 is an example of a galaxy viewed edge-on from Earth. Photo by Tim Hunter.

O deep of Heaven, 't is thou alone art boundless,
'T is thou alone our balance shall not weigh,
'T is thou alone our fathom-line finds soundless,—
Whose infinite our finite must obey!
Through thy blue realms and down thy starry reaches
Thought voyages forth beyond the furthest fire,
And, homing from no sighted shoreline, teaches
Thee measureless as in the soul's desire.
O deep of Heaven, no beam of Pleiad ranging
Eternity may bridge thy gulf of spheres!
The ceaseless hum that fills thy sleep unchanging
Is rain of the innumerable years.
Our worlds, our suns, our ages, these but stream
Through thine abiding like a dateless dream.

—Sir Charles G. D. Roberts, "The Night Sky" (1889), The Collected Poems of Sir Charles G. D. Roberts, ed. Desmond Pacey (Wolfville, N.S.: Wombat, 1985), p. 120.

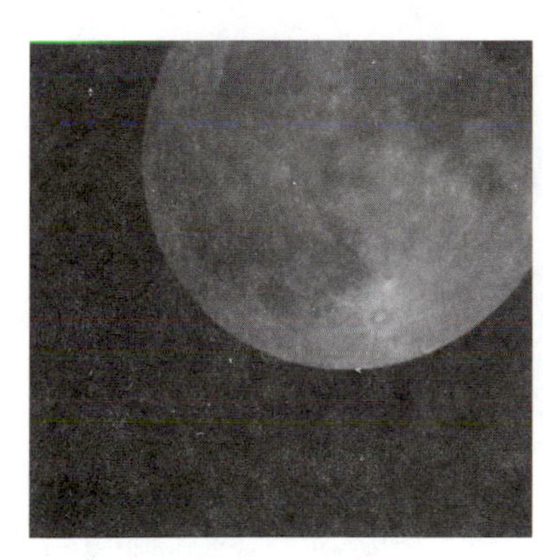

The Moon. Photo by David Levy.

Above: Scorpius-Sagittarius Milky Way. Photo by Roy Bishop.

I wandered lonely as a cloud
That floats on high o'er vales and hills,
When all at once I saw a crowd,
A host, of golden daffodils;
Beside the lake, beneath the trees,
Fluttering and dancing in the breeze.

Continuous as the stars that shine
And twinkle on the milky way,
They stretched in never-ending line
Along the margin of a bay:
Ten thousand saw I at a glance,
Tossing their heads in sprightly dance.

—William Wordsworth, "I Wandered Lonely as a Cloud" (1804), English Romantic Poetry and Prose, ed. Russell Noyes (New York: Oxford University Press, 1956), p. 325.

I will lift up mine eyes unto the mountains:
From whence cometh my help . . .
—Psalm 121:1

Twilight at Palomar Mountain Observatory. This view shows the dome of the two hundred-inch telescope taken through the open shutters of the 18-inch telescope dome. Photo by David Levy.

Chapter One

Nightfall

N*ightfall is a* magic time. As the blue sky of day settles into a navy blue and then blue-black of night, the curtain of heaven rises, showing stars, planets, and the Milky Way straddling the sky. With its promise of peace and repose, night beckons us, whether we are astronomers, poets, painters, or just people enjoying an evening stroll. It courted poets like Thomas Gray, whose words, written in the 1740s, can turn a simple scene of a darkening sky into one of the most heartfelt images in the English language:

The curfew tolls the knell of parting day,
The lowing herd wind slowly o'er the lea,
The plowman homeward plods his weary way,
And leaves the world to darkness, and to me.

Now fades the glimmering landscape on the sight,
And all the air a solemn stillness holds,
Save where the beetle wheels his droning flight,
And drowsy tinklings lull the distant folds;

Save that from yonder ivy-mantled tower
The moping owl does to the moon complain
Of such, as wand'ring near her secret bower
Molest her ancient solitary reign.[1]

One hundred forty years after Gray wrote these words, Vincent Van Gogh was also struck by the magnificence of night. His resulting astronomical art is brilliant almost beyond description. He painted a nightfall that including the Moon, and probably Venus and Mercury, in *Road with Cypress and Star*. Two of his works of art have "Starry Night" in their titles. The earlier one, completed in the autumn of 1888, was *Starry Night over the River Rhône*. Although the landscape is viewed to the southwest, the seven stars of the Big Dipper are clearly portrayed as seen to the north; artistic license won out over scientific accuracy, but the pleasing result shows the bowl of the dipper seemingly gathering water from the river.

Then there is *Starry Night,* one of the most celebrated works of art in existence. I have heard that the stars Van Gogh depicts variously refer to Venus and the Moon, which were in the morning sky at the end of June 1889, the month he painted it. More intriguing than the stars and Moon, however, are the swirls of hazy starlight that cross the sky over the town. On a first look, these swirls, and the stars that they surround and cross, seem to be ordinary stars seen through the eyes of someone intoxicated or extremely nearsighted. The artist, however, might have had much more in mind. To understand this, we need to travel in space and in time from St. Rémy, where the artist lived and saw the stars he painted, to 1850 and the town of Birr, Ireland, where amateur astronomer Lord Rosse was observing through the world's largest telescope, a reflector with a mirror seven feet wide.

With the mirror end on the ground, the mighty telescope's long bulky tube was slung with ropes between two large brick walls. Using a staircase to reaching his eyepiece, Rosse observed and drew all kinds of celestial objects, from fields of stars to those misty blobs of cloudy light called nebulae. Like other astronomers of his day, Rosse could only speculate as to what these spiral-shaped nebulae might be; solar systems in the process of formation was the most popular idea.

Messier 51, the Whirlpool, was one of these objects. Located just south of the Big Dipper's handle, this distant maelstrom was discovered on October 13, 1773, by the famous French comet discoverer Charles Messier and listed as No. 51 in his catalogue of objects that could be mistaken for comets. At the time, Messier 51 was thought to be a solar system in formation. In 1903, V. M. Slipher, an astronomer at Lowell Observatory in Flagstaff, Arizona, took a series of long exposure photographs of several of these spiral nebulae. Using the slow films of the time, Slipher had to spend as much as two full nights exposing his glass plate to gather enough light to obtain a spectrum of a single nebula. The results were uniformly strange: all the spectra of the spiral nebulae, when compared to the spectrum of our Sun, were shifted toward the red end of the spectrum. Slipher did not know why this should happen, as planets in formation around Sun-like stars should have spectra similar to that of the Sun. It was not until 1924 that Edwin Hubble solved the mystery. Using the great 100-inch telescope atop Mt. Wilson, he and Milton Humason took photographs that were sharp enough to resolve these hazy spots into huge emporiums of stars.

Messier 51 is now known to be a vast spiral galaxy some fifteen million light years away. Informally called the Whirlpool Galaxy, M51 is fifty thousand light years across, and, rather than one Sun about to have planets, this galaxy shines with the intensity of ten billion suns.[2] It can be spotted as a diffuse patch of light through small telescopes, but through large telescopes the vastness of the spiral structure is spectacular.

Some years ago Wendee, my wife, and I explored the details of M51 with the sixty-one-inch Kuiper telescope in the Catalina north of Tucson. The galaxy seemed to have no end as it spiraled outward from its center. It was a staggering sight, unchanged from the beauty that Rosse had seen so long ago. It was the first time Wendee actually saw the separate arms unfolding from the galaxy's center, and we were both entranced.

In May of 1889, Vincent Van Gogh entered the asylum of St.-Paul-de-Mausole, located near the French town of St. Rémy. The asylum was caring for only ten male patients at the time, so Van Gogh had the luxury of two rooms, one for himself and the other as his studio. By the end of May, Van Gogh seemed at peace there, and painted some of the most magnificent of his later landscape paintings. In June he painted *Starry Night*, "an indisputably visionary painting," writes critic Nathaniel Harris, "in which a huge moon, radiant stars or suns and swirling, wave-like nebulae fill the night sky. A cypress and the entire landscape tremble and move in sympathy; only the little village remains solid and stolid in the turbulent vastness of the universe."[3]

Perhaps, as Hillary Clinton might say, it takes a village to pacify the universe. The village in Van Gogh's painting contains peaceful, sleepy structures outlined with straight lines. With everyone presumably asleep in the village, there seems no relation between it and the sky except for the church, whose straight spire reaches toward the sky. Steeple meets sky in the midst of the outermost nebula, as though it is asking for some clarification on why the sky seems so out of phase. The town of St. Rémy was not a friendly place to Van Gogh: "I fell ill," he said a month after the painting, "from the mere sight of people and things.[4] *Starry Night* seems to depict its artist's preference for disorder and unbridled energy over the ordered quiet of the village.

In my own college years I went through a period of depression that required hospitalization and a lot of therapy to cure. During those years, during the times I was relaxed enough to appreciate it, the peace and permanence of the night sky offered a sense of stability. "It is now three o'clock in the morning," I wrote on May 22, 1973, my twenty-fifth birthday, "and I have just completed an observing session that couldn't have lasted half an hour, yet turned out to be a fine communion with a part of Nature which I have always loved, but in these last months have ignored. I realize tonight that it does not matter whether I hunt for comets, or obtain magnitude estimates of variable stars, or stay out all night. The good observing session means a private feeling of a successful rendezvous with Vega or Jupiter, as in tonight's case, or Saturn and Sirius and Canopus on another night."[5]

Painting of a Whirlpool

In 1985, Charles Whitney, an astronomer at Harvard, pointed out the similarity of the nebulous swirls to Lord Rosse's famous 1845 drawing of Messier 51. His research complemented that of art historian Albert Boime, who noted that Van Gogh was familiar with the sketch, having studied it, probably, from a copy of *Les Etoiles*, a book by famous astronomy writer

Camille Flammarion that included the Lord Rosse drawing. It is possible that Van Gogh even noticed the book's front page illustration of a church spire pointing to the sky.[6]

When one compares the swirls of *Starry Night* with Lord Rosse's drawing, they seem to match better if the painting is reversed; the spiral shape in M51 seems counterclockwise, though the artist made it clockwise. He might have felt free to make similar reversals in his *Road with Cypress and Star*, painted in April 1890. On the evening of April 20, the Moon was in a beautiful grouping with Venus and Mercury; however, if the painting indeed does include these two planets, the grouping is reversed.[7] Weather records studied by Charles Whitney indicate that the evening of April 20 was the first cloud-free night in a week, making it more likely that Van Gogh would have been motivated to go outdoors to paint.[8] Another suggestion: Van Gogh faced eastward, away from the twilight sky, so that whatever daylight was left would have illuminated his canvas. If he then used a mirror to see the Moon and planets, it would have reversed their appearance relative to one another.[9]

None of these details or speculations diminishes the wonder of the artist's communion with the forces of nature. "There is certainly an affinity between a person and his work," Van Gogh told his brother Theo, "but it is not easy to define what that affinity is."[10] For *Starry Night*, Van Gogh confronts his universe with celestial objects of marked contrast; the sharp, almost violent images of Moon and stars, the silent dagger of the cypress tree, all opposed by the graceful swirls of the Whirlpool.

Flashing across the Sky

Imagine a comet wandering through space like an ancient sailing ship. Rounding the Sun every few hundred thousand years, after a near miss with Jupiter, it changes its path and visits the Sun more frequently. Eventually its elliptical journey is shortened so greatly that it now orbits the Sun every seventy-six years. On one pass, a tiny particle of dust escapes. This particle travels on its own through space, rounding the Sun and occasionally dashing past Earth. Finally, one clear October night this particle, now racing along at some forty miles per second, hurtles through the Earth's atmosphere. In a second it vaporizes to nothing, just as an eight-year-old looks up and watches. The speck of dust has given its 4.5 billion year life to ignite a child's interest in the sky.

We have all seen shooting stars, which are not stars at all but the glowing air surrounding tiny particles that vaporize in the upper atmosphere of Earth. They look like falling stars, yet after one appears, the number of stars in the sky has not decreased. It is the bright flash of ionizing gases around the particle that we call a meteor. If the particle is larger, say, than the size of a tennis ball, it would hit Earth's atmosphere with a very bright streak. If it is as large as a basketball, it might survive its fall, leaving small pieces to hit the ground. These pieces are called meteorites.

In 1923, American poet Robert Frost published a poem about a small meteorite, a tiny remnant of a body in space, that ended up as part of a wall:

> *Never tell me that not one star of all*
> *That slip from heaven at night and softly fall*
> *Has been picked up with stones to build a wall.*
>
> *Some laborer found one faded and stone cold.*[11]

These lines beckon us to envision the intricate relationship among the sky, the Earth, and those who inhabit the Earth. We see a wall that stands as part of a house, or as part of a fence, or as the remains of the ancient Temple in Jerusalem—an artificial wall built with stones and bonding materials for some purpose. All of its stones, while ancient and enduring, have been gathered from some local quarry. But one stone in one wall is different. Traveling through space and hurtling to Earth, this rock, with so much to tell, has simply taken its place as one of many to hold up a wall, an ignominious end for such a noble stone.

Meteors are such simple parts of nature, and yet so wonderful, so mysterious; their beauty transcends their brevity, and they have captured the pens of earlier poets. In the fifteenth century John Donne began a song about explaining wonderful things:

Goe, and catche a falling starre,
Get with child a mandrake roote,
Tell me, where all past yeares are,
Or who cleft the Devil's foot.
Teach me to heare Mermaides singing,
Or to keep off envies stinging,
And find
What winde
Serves to advance an honest minde.[12]

Two hundred years later Byron pondered:

When the moon is on the wave,
And the glow-worm in the grass,
And the meteor on the grave,
And the wisp on the morass;
When the falling stars are shooting,
And the answer'd owls are hooting,
And the silent leaves are still
In the shadow of the hill
Shall my soul be upon thine,
With a power and with a sign.[13]

Although these references to meteors appear almost incidental, they seem to share the notion that there is something momentous in the fall of matter from the sky. Robert Southey wrote at the start of the 19th century in "St. Antidius, the Pope and the Devil":

He ran against a shooting star,
So fast for fear did he sail,
And he singed the beard of the Bishop
Against a comet's tail;
And he passed between the horns of the moon,
With Antidius on his back;
And there was an eclipse that night,
Which was not in the Almanac.[14]

The Moon

The Moon as a marker of time and space is one of my favorite images, a picture often conjured in English poetry. It is Robert Frost's natural timepiece: Frost referred to the Moon as a "luminary clock" that "proclaimed the time was neither wrong nor right."[15]

Here the Moon appears in Coleridge's "Rime of the Ancient Mariner":

The moving Moon went up the sky,
And no where did abide:
Softly she was going up,
And a star or two beside—[16]

Although most of us never realize it, the Moon does keep its own time, rising thirty minutes to an hour later each night. On fully half the nights of the month, the Moon either rises so early that it is visible during the afternoon, or so late that it stays in the sky until late in the morning. Hopkins noted this in his 1862 piece "A fragment of anything you like":

Fair, but of fairness as a vision dream'd;
Dry were her sad eyes that would fain have stream'd;
She stood before a light not hers, and seem'd

The lorn Moon, pale with piteous dismay,
Who rising late had miss'd her painful way
In wandering until broad light of day;

Then was discover'd in the pathless sky,
White-faced, as one in sad assay to fly
Who asks not life but only place to die.[17]

The appearance of the Moon—not just its timing, but its shape, and its effect on clouds and landscape—has also long been a poetic image of choice. Whether the Moon is in the sky at night is a question that interests poets and painters as well as astronomers. A moonlit night welcomes peaceful evening walks that are less inviting when the night sky is dark. However, it is only during the four days nearest full Moon when moonlight makes a very noticeable difference—only two and a half days before full Moon the sky is only half as bright. Thomas Gray could turn the absence of moonlight into a profound sense of loss, as seen in this passage from a journal Gray kept of a tour through English Lakes:

> At distance heard the murmur of many waterfalls not audible in the day-time. Wish'd for the Moon, but she was dark to me & silent, hid in her vacant interlunar cave.[18]

Gray quoted the last six words in this passage from Milton's *Samson Agonistes*. The blind Samson says:

> *The Sun to me is dark*
> *And silent as the Moon,*
> *When she deserts the night,*
> *Hid in her vacant interlunar cave.*

When the Moon shines through high cirrus clouds, it often surrounds itself with a halo of light.

Samuel Taylor Coleridge wrote:

> The Moon, rushing onward through the coursing clouds, advances like an indignant warrior through a fleeing army; but the amber halo in which he moves—O! It is a circle of Hope. For what she leaves behind her has not lost its radiance as it is melting away into oblivion, while, still, the other semicircle catches the rich light at her approach, and heralds her ongress.[19]

Gerard Manley Hopkins wrote often of haloes, like this lunar one of February 1872:

> The halo was not quite round, for in the first place it was a little pulled and drawn below, by the refraction of the lower air perhaps, but what is more it fell in on the nether left hand side to rhyme the moon itself, which was not quite at full. I could not but strongly feel in my fancy the odd instress of this, the moon leaning on her side, as if fallen back, in the cheerful light floor within the ring, after with magical rightness and success tracing round her the ring the steady copy of her own outline.[20]

Whenever the Moon's phase is thin enough, we can see the dark part as well as the bright. This effect, called earthshine, results from the Earth's shining on the Moon's dark side. Anyone standing on the dark part of the Moon at such a time would behold a sky brightly lit by the Earth. Here at home, earthshine is not to be feared, despite this poor sailor's feeling in "The Ballad of Sir Patrick Spence":

Late, late yestreen I saw the new moone
Wi' the auld moone in her arme;
And I feir, I feir, my deir master,
That we will com to harme.[21]

Nor can a star appear within the dark section of the Moon, notwithstanding this bad omen from Coleridge's "Rime of the Ancient Mariner":

We listened and looked sideways up!
Fear at my heart, as at a cup,
My lifeblood seemed to sip!
The stars were dim, and thick the night,
The steersman's face by his lamp gleamed white:
From the sails the dew did drip—
Till clomb above the eastern bar
The hornèd Moon, with one bright star
Within the nether tip.[22]

The Moon has an aspect that stands beyond time, though, or at least beyond the length of time we can imagine: the story of its cratered surface we see through binoculars or a telescope. While here on Earth we live in areas surrounded by local rocks of a span of ages, a single look at the Moon gives the whole panorama of its 4.5 billion year history. The Moon was probably created as the result of a shotgun marriage between two great worlds, the Earth and another large body with which it collided more than four billion years ago. So violent was the crash that the Earth's crust melted away, and a storm of rocky material from the Earth and the other world formed a thick ring around the Earth. Eventually that material congealed to form the Moon. And if the Moon was born in violence, it has lived its life that way, too. Even without a telescope or binoculars, anyone can see the large grey areas that form the face of what some perceive as "the man in the Moon." Some of these areas are impact basins, formed when objects struck the Moon in an ancient blitz that took place some 3.9 billion years ago. Much later these basins filled with dark basalt, a lava that gives them their dark appearance. Finally, through a telescope, one can see hundreds of small craters, all formed when asteroids or comets struck the Moon throughout its history. One of the youngest major craters on the Moon, Tycho, was formed when a comet or an asteroid struck the Moon some 100 million years ago. When the Moon is near its full phase, one can see the rays of rocky material that pan out from the center of that impact site, marking this relatively recent change in the Moon's long history.

> *Think'st thou I'd make a life of jealousy;*
> *To follow still the changes of the moon*
> *With fresh suspicions? No; to be once in doubt*
> *Is once to be resolved.*
>
> Shakespeare, *Othello*[23]

Eclipses

When the Earth, Moon, and Sun align perfectly, the shadow of one falls on another and an eclipse occurs. A solar eclipse occurs when the Moon is positioned directly between the Earth and Sun, and its shadow falls on a small portion of the Earth. A lunar eclipse occurs when the Earth travels between the Sun and the Moon; the darkening of the Moon is the effect of the Moon's passing directly through the Earth's shadow. Evoking feelings of wonder, eclipses are not everyday events. Not only was Shakespeare well aware of the beauty and power of an eclipse, but he also understood the debate over the role that these events play in our lives. In *King Lear* he constructed the famous scene between the Earl of Gloucester and Edmund, his bastard son, that illustrates their different views about whether an astronomical event can portend evil for humanity. In Edmund's opening appearance he announces his commitment to rational thinking: "Thou, Nature, art my goddess; to thy Law /My services are bound."[24] Still he knows that nature does not rule our thoughts; we do. In contrast, in a speech permeated with concern for the future, The Earl of Gloucester argues: "These late eclipses in the sun and moon portend no good to us. Though the wisdom of nature can reason it thus and thus, yet nature finds itself scourged by the sequent effects. Love cools, friendship falls off, brothers divide."[25]

After Gloucester storms out, Edmund laughs: "This is the excellent foppery of the world, that when we are sick in fortune, often the surfeits of our own behavior, we make guilty of our disasters the sun, the moon, and stars; as if we were villains on necessity."[26]

Anyone who has experienced the grandeur of an eclipse, be it of the Sun or the Moon, understands the poetic inspiration that it evokes. Eclipses are profound events, in no small part because their four-body lineup is one in which we participate—Sun, Moon, Earth, and observer. In a sense, we observers are not unlike the meteorite that ended up in a wall in Frost's poem quoted earlier in this chapter. In recent years we have come to understand that life itself might have begun as a result of organic materials being delivered to Earth as a result of impacts of comets. But in our ability to appreciate that relation, we differ from that meteor. In July 1991, I watched a magnificent total solar eclipse from Mexico's Baja California peninsula.

The sky did not darken linearly and steadily, as it does after sunset. For the first half hour, the bright summer sky made it hard to imagine that anything unusual was happening at all. My first thought that the light was changing came as I noticed that the sky was not quite as bright as it should be for a summer desert near noon. Through my welder's glass I looked and saw that fully half the Sun was gone. By the time the Sun was three-quarters covered, the pace of darkening was increasing rapidly. Now navy blue, the sky was darkening as fast as if someone were operating a dimmer switch. Soon there was just a sliver of Sun left, and I could see the Sun shrink as the seconds ticked away. Directly to the west the dark shadow of the Moon gained substance. The scene ripped to the core of my being; something inside wanted to hold it back for a few minutes so that I could find a key and lock this moment. But still the sky grew darker.

When I looked again, the Sun had gone. In its place was a golden crown stretching some three solar diameters east and west; through the telescope this garland was rich with streamers and intricate brushes of light. On the north and south sides were a series of smaller eruptions of rays, shining outward like the mouths of baby birds in a nest.

Eclipses heighten the senses: listening to the sounds of the eclipse—the birds flying and strutting about, the incessant clicks of a million cameras—all accompanied the eerie light that surrounded us. Around the horizon was a bright red glow. In the final seconds of totality the western sky started to brighten rapidly. I looked back at the Sun, admiring a red tongue of flame, or prominence, as the western edge

The total eclipse of the Sun, 26 February 1969. Photo by David Levy.

brightened quickly. Suddenly a sharp speck of photosphere stabbed the darkness, then slowly spread out into a thin crescent. The eclipse was over, and yet it wasn't. As the Moon crept away, it left an ever larger crescent Sun, a rapidly brightening landscape, and a euphoric crowd of people. No one doubted that an event like this had the potential to terrify people long ago.

Milton also understood the strong, basic feelings that solar eclipses generate. He wrote of solar eclipses in *Samson Agonistes*, in which the blinded Samson laments his betrayal and loss of sight, comparing his blindness to a total eclipse:

> *O dark, dark, dark, amid the blaze of noon,*
> *Irrecoverably dark, total eclipse*
> *Without all hope of day!*[27]

In *Paradise Lost*, he wrote that the Sun

> *In dim eclipse, disastrous twilight sheds*
> *On half the nations, and with fear of change*
> *Perplexes monarchs.*[28]

Some three centuries after *Paradise Lost*, the great novelist Thomas Hardy was inspired to write as Milton did about shadows and nations. But Hardy's eclipse was of the Moon, not the Sun, and while a lunar eclipse lacks the drama of a solar eclipse, it offers its own special majesty to anyone who views it. Spying a coppery red Moon in the sky offers a feeling of serenity more than of drama. "At a Lunar Eclipse," written in 1903, contrasts the apparent peace of Earth's shadow with an Earth in turmoil:

Thy shadow, Earth, from Pole to Central Sea,
Now steals along upon the Moon's meek shine
In even monochrome and curving line
Of imperturbable serenity.

How shall I link such sun-cast symmetry
With the torn troubled form I know as thine,
That profile, placid as a brow divine,
With continents of moil and misery?

And can immense mortality but throw
So small a shade, and Heaven's high human scheme
Be hemmed within the coasts yon arc implies?

Is such a stellar guage of earthly show,
Nation at war with nation, brains that teem,
Heroes, and women fairer than the skies?[29]

When the eclipse is over, we are left with what we've always had—a clear night aglow with stars. If the upper atmospheric winds are strong, the stars will appear to twinkle, apparently sending out alternating colors. These have nothing to do with the star's actual color, just with a testimonial to the final millionth of a second of the journey of the star's light from the star to Earth. In 1855 Robert Browning, in his poem "My Star," wrote of this effect:

All that I know
Of a certain star,
Is, can it throw
(Like the angled spar)
Now a dart of red,
Now a dart of blue;
Till my friends have said
They would fain see, too,
My star that dartles the red and the blue!
Then it stops like a bird; like a flower, hangs furled:
They must solace themselves with the Saturn above it.
What matter to me if their star is a world?
Mine has opened its soul to me; therefore I love it.[30]

This overwhelming feeling of love for a star, even more the whole sky, is something with which forty years of nightly observing have made me very familiar. The stars beckon to us, and even open, as Browning writes, their "souls" to us. But as Mary Lozer, a contemporary Arizona poet, points out in her 1999 poem "Thirsting," sharing one's feelings with a star is not an experience that happens overnight:

Star light, star bright,
Which shall capture my attention tonight?
As a fickle lover,
My eyes glance from one to another,
Resting a moment on the familiar,
Moving on to striking my fancy.
Like a roving dandy, I want it all.

Patience, patience, they seem to implore.
Stay awhile; lengthen your visit,
We are Scheherazade's,
With many stories to tell.

Take more than a moment,
What is your hurry, Earth Dweller?
Our stay is so long,
Yours is so short,
Is that why you gulp us into yourself
Like a thirsty wanderer?[31]

Bright Star

Van Gogh's swirling sky seemes, ironically, to offer peace, or at least an alternative to a troublesome life. Is the artist calling to the sky that he paints in *Starry Night,* just as Keats called, forty years earlier in his 1819 sonnet to Fanny Brawne, to a lone bright star?

Bright star, would I were stedfast as thou art—
Not in lone splendor hung aloft the night
And watching, with eternal lids apart,
Like nature's patient, sleepless Eremite,
The moving waters at their priestlike task
Of pure ablution round earth's human shores,
Or gazing on the new soft fallen mask
Of snow upon the mountains and the moors—
No—yet still stedfast, still unchangeable,
Pillowed upon my fair love's ripening breast,
To feel forever its soft fall and swell,
Awake forever in a sweet unrest,
Still still to hear her tender-taken breath,
And so live ever—or else swoon to death.[32]

Calling to the heavens can produce an unintended result, as Stephen Crane wrote in 1899, a year before his death:

A man said to the universe,
"Sir, I exist!"
"However," replied the universe,
"The fact has not created in me
A sense of obligation."[33]

This marvellous little poem features an insignificant man trying to comfort the universe with its responsibility for human existence.

Not so, wrote Max Ehrmann in 1948's "Desiderata": "You are a child of the universe, no less than the trees and the stars, you have a right to be here. And whether or not it is clear to you, no doubt the universe is unfolding as it should."[34]

When a comet sails through the night, or an eclipse brings darkness at noon, we look toward the heavens with a sense of wonder, sensing that we are a part of the Universe. We pose the same question that Robert Browning asked in 1855, in "Andrea del Sarto":

> *Ah, but a man's reach should exceed his grasp,*
> *Or what's a heaven for?*[35]

These events in the night sky are magical, their "imperturbable serenity" inspiring us to rise above who we are. They link us to times long past, when humans first looked up in wonder as a meteor fell from the sky. When I am setting up my telescope as evening falls, I might feel as if 250 years of time have vanished, and I am sitting beside Thomas Gray as the parting day leaves his world to darkness. Hours later, as I search for comets in the blissful quiet before dawn, and my telescope happens to pass over Messier 51, the Whirlpool Galaxy, I am next to Van Gogh as he looks out to the heavens, paints the stars on canvas, then adds the swirls of the Whirlpool's distant suns to the majesty of his night sky.

Chapter Two

Stars, Hide Your Fires

The universe is a most fertile subject for poetry, and considering the importance of the sky to early watchers, it is not surprising to find astronomical references even in ancient literature. The first book of Chronicles describes what could be a comet—the comet of 971 B.C.E. appeared near that time—which according to 1 Chronicles signaled God's protest of an ill-advised census King David had ordered. The biblical passage is read every year at the Passover seder: *And David lifted up his eyes, and saw the angel of the Lord standing between the earth and the heaven, having a drawn sword in his hand stretched out over Jerusalem.*[1]

Around 760 B.C.E., the prophet Amos wrote of what could have been the solar eclipse of 762 B.C.E. in Ninevah:

And it shall come to pass in that day,
Saith the Lord God,
That I will cause the sun to go down at noon
And I will darken the earth in the clear day.[2]

In Isaiah there is a strange passage that may refer to a large partial solar eclipse over Jerusalem on January 11, 688 B.C.E.:

> Behold, I will cause the shadow of the dial, which is gone down on the sun-dial of Ahaz, to return backward ten degrees. So the sun returned ten degrees, by which degrees it was gone down.[3]

Medieval English Literature Before the New Philosophy

In the fourteenth century, Geoffrey Chaucer's *Treatise on the Astrolabe* was a rare literary work which explained the workings of the major astronomical instrument of the time, the astrolabe—a two-dimensional model of the heavens used for determining positions of stars and planets, as well as times of rising and setting of specific objects. This was as surprising a production as if the most famous poet of our age should write a treatise about modern astronomical telescopes. It is possible that Chaucer took a much dimmer view of using the stars to predict one's future than he did of their scientific value. In *The Miller's Tale,* one of the most humorous of the *Canterbury Tales,* Geoffrey Chaucer pokes fun at the astrologer who walks into a field to gaze at the stars to discern what they predict, and falls into a pit:

So ferde another clerk with astromye;
He walked into the feeldes, for to prye
Upon the sterres, what ther sholde bifalle,
Til he was in a marle-pit yfalle:
He saugh nat that.[4]

Of all the objects in the sky, the temporary ones—the exploding stars or novae—and the moving ones—the comets—obviously attracted the most attention in ancient times. Rarely a supernova would blast forth, a star whose core collapses in a titanic explosion. In historic times one appeared in 1054, a second in 1572, and a third in 1604. Of the moving objects, comets flew in and then moved away, bringing with them a xenophobic fear of the unknown.

In *Julius Caesar*, Shakespeare writes of Calpurnia, who, having dreamt of an apparition in the sky, begged her husband Julius Caesar not to travel to the Senate: "When beggars die," Calpurnia exclaimed, "there are no comets seen. The heavens themselves blaze forth the death of princes."[5] Caesar ignored her pleas, left for the Senate anyway, and was assassinated as he sat under Pompeii's statue. *"Et tu, Brute!"* he moaned as he saw his friend Brutus leading the plot, "Then fall, Caesar!"

Calpurnia might have been right about the comet, but the historical record tells us that she could not have seen one before her husband was killed. During the Octavian games held a few weeks after his death, a bright comet moved slowly from night to night in the northern sky. The comet was widely interpreted as Caesar's soul traveling to the stars.

More Things in Heaven and Earth: Astrological Myth and Science in Shakespeare

Shakespeare's time coincided with a number of unusual events, especially the suggestion by Copernicus as far back as 1543 that the Earth was not the center of the universe. However, at the time of most of Shakespeare's writing, the real impact of Copernicus' ideas had yet to be felt. Cosmic events were happening, but slowly. There was a supernova in 1572; the great astronomer Tycho Brahe interpreted it as meaning that the sphere of fixed stars that lay beyond the planets might not be so unchangeable as people had thought. Five years later, a bright comet led Tycho to suggest that comets were not a part of the atmosphere (a view that had dominated cometary science since Aristotle first suggested it) and that they moved methodically in a sphere beyond the Moon. Queen Elizabeth herself defied royal tradition not to set eyes on transient events in the sky by observing the comet of 1582, and a second supernova appeared in 1604.

Tycho's ideas were based on meticulous observation, rather than the theoretical arguments that led philosophers to accept the *Almagest* of Ptolemy, a work that championed the Earth-centered universe and which was considered as virtual gospel since its appearance around 150 C.E. Far in advance of his time, Tycho's idea of basing arguments on observation would not gain support until 1620, when Francis Bacon published *Novum Organum* (see next chapter).

Shakespeare's references to the sky are rich in the casual astrology of the time, as when, for example, the two characters Helena and Parolles, in *All's Well That Ends Well*, trade views on their astrological predicaments:

> *Helena:* *Monsieur Parolles, you were born under a charitable star.*
> *Parolles:* *Under Mars, I.*
> *Helena:* *I especially think, under Mars.*
> *Parolles:* *Why under Mars?*
> *Helena:* *The wars hath kept you under that you must needs be born under Mars.*
> *Parolles:* *When he was predominant.*
> *Helena:* *When he was retrograde, I think, rather.*
> *Parolles:* *Why think you so?*
> *Helena:* *You go so much backward when you fight.*
> *Parolles:* *That's for advantage.*
> *Helena:* *So is running away, when fear proposes the safety; but the composition that your valour and fear makes in you is a virtue of a good wing, and I like the wear well.*[6]

Had Shakespeare's prime writing years ended a decade later, his plays might have reflected a vastly different situation. After Galileo's crucial observations of Jupiter, Venus, the Moon, and the Sun, philosophers and writers had to contend with the mounting evidence for the new philosophy of a universe in which the Earth circles the Sun rather than one in which the Sun orbits the Earth. References to the Sun-centered universe in writing before 1610 are rare, but after that critical year they become more common. Shakespeare, according to the well-known critic Marjorie Hope Nicholson, "lived in a world of Time, Milton in a universe of Space. It happened as suddenly as that."[7]

In chapter one we explored Gloucester's speech about the "late eclipses" of both Sun and Moon. These eclipses were actually used to help date the play, but two eclipses in 1605 that are most commonly cited may not have been the ones Shakespeare had in mind. Johnstone Parr, among other critics, wondered which of the three solar eclipses that occurred around the writing of *Lear*, or whether all had been planted in Gloucester's thoughts. But the historical record shows that England was free of *any* major solar eclipse for more than half a century, a most unusually long period that lasted from 1547 to 1598. Then three major eclipses of the Sun tracked close to England in rapid succession. The central line of the annular eclipse of 1601, in fact, passed near London. In such an eclipse, the Moon is too far from Earth to cover the Sun completely as it passes in front of it, so that viewers on the central line would see an annulus, or ring, of sunlight surrounding the darkened Moon.

Shakespeare may well have had the September 27, 1605, eclipse of the Sun, as well as the lunar eclipse that preceded it by two weeks, in mind when he wrote Gloucester's words in *Lear*. But the solar eclipse was a small one that would not have created great excitement. More likely, Shakespeare intended a general reference to the sudden occurrence of several eclipses.[8] There are other references to eclipses such as the poignant one in *Antony and Cleopatra*, where Anthony fears:

Alack, our terrene moon
Is now eclips'd, and it portends alone
The fall of Antony.[9]

These references are to astronomical events, which would have interested Shakespeare's audiences more than arcane discussions of astronomical theory would. In fact, we find very few Shakespearean references to Copernicus and his ideas. Hamlet's love letter to Ophelia, however, may be an exception:

Doubt thou the stars are fire,
Doubt that the sun doth move,
Doubt truth to be a liar,
But never doubt I love.[10]

Although there are several versions of Hamlet, the line "doubt that the sun doth move" does appear in the first good text of the play, the second quarto published around 1604 or 1605. This was more than half a decade before Galileo's observations. Even if Shakespeare had believed in the new cosmology, it would not have served his purpose well, for the old system, with its emphasis on the Earth and mankind at the center of the universe, is more sound for the purpose of drama. After all, why would the heavens have blazed forth the death of princes in Calpurnia's dream, if the princes had not been at the center of the universe?

What is interesting is not which set of cosmic rules concerned Shakespeare but that he did have a lively interest in the sky. This interest is obvious in the 135 references in the canon to stars, 19 to planets, 11 to eclipses, and seven to comets,[11] not to mention the 800 less specific invocations to heaven of which the most famous is probably Hamlet's cry to his friend:

There are more things in heaven and earth, Horatio,
Than are dreamt of in your philosophy.

Besides this listing is the fact, noted by critic Wolfgang Clemen, that almost every major tragic hero in Shakespeare must contend with the forces of the cosmos, a trait not matched by the heroes in the histories or the comedies:

> When in the histories, the people turned their eyes to the sun, taking its dull gleam for a foreboding of evil, this was in the tradition of omen. But in the tragedies, the characters apostrophize the sun and stars directly. "Stars, hide your fires, . . ." (I: iv, 50) Macbeth cries before his murderous deed. "Moon and stars!" we hear Antony say, and Cleopatra: "O Sun, burn the great sphere. . . ." The heavens seem sympathetic to what is occurring here on earth. To Hamlet, thinking of his mother's hasty remarriage, "heaven's face doth glow" (III: iv, 48) and Othello, convinced of Desdemona's faithlessness, cries out: "Heaven stops the nose at it and the moon winks" (IV: ii, 77). . . . Moreover, in the dramatic structure of the individual tragedies the appeal to the elements makes its appearance at definite turning-points. Not until they begin to despair of men and earth do the tragic heroes turn to the heavens. When their firmest beliefs have been shaken, when they stand alone and forsaken, they renounce the earth and call upon the cosmic powers.[12]

The single concept that all these apostrophes have in common is not a fear but a respect of the power of the cosmos. In Renaissance England the cosmic debate involved two separate but related ideas, both called astrology. There was the *natural astrology*, which considered the planetary motions and the positions of other celestial objects in their spheres, and the *judicial astrology*, which concerned itself with the effects of these motions on humans. The modern term "astronomy" was, in Elizabethan times, used as a synonym for either branch of astrology, as Edgar shows us in *Lear*:

> *Edmund:* *I am thinking, brother, of a prediction I read the other day what should follow these eclipses.*
> *Edgar:* *Do you busy yourself with that? . . .*
> *How long have you been a sectary astronomical?*[13]

There is considerable debate as to how seriously many Elizabethans followed or believed in the notions and predictions of the judicial astrologers. Critic J. W. Draper suggests that the people were committed to these ideas, and that Queen Elizabeth regularly employed the learned Dr. Dee to compute for her the lucky days and hours for conducting her affairs.[14] However, Elizabeth defied her own council by observing the Great Comet of 1582, and Dr. Dee was known as a Copernican astrologer whose advice was received more for its informative value than for its predictive power. In any event, at the time, there was considerable religious opposition to judicial astrology, since its practice violated the first commandment. According to Warren D. Smith, the fact that Elizabeth's second parliament passed the first of a series of antijudicial astrology bills, each harsher than the last, indicates that people did practice astrology and that other people objected to its practice. The Act of 1580, in fact, compared judicial astrology to witchcraft and threatened the death penalty for the practice of either. James I also disdained the practice, distinguishing between *astronomia,* the law of the stars, and *astrologia,* the preaching of the stars. Of the latter James wrote: "It is this part which I called before the devils schole."[15]

If Elizabethans, or at least their government, scoffed at the judicial astrologers, they certainly enjoyed hearing from them throughout Shakespeare's plays. We do not have to look far to find evidence that Shakespeare capitalized on this popular interest from his earliest plays to his last. In *The Two Gentlemen of Verona,* Julia speaks of her love for Proteus:

> *Base men, that use them to so base effect!*
> *But truer stars did govern Proteus' birth;*
> *His words are bonds. . . .*[16]

We have already seen Helena praise Parolles' "charitable star" in *All's Well That Ends Well.* It appears that Shakespeare was familiar with astrological concepts, and used them effectively. *Richard II,* for example, offers a far more serious invocation to the cosmos' judicial power:

> *Captain:* *'Tis thought the king is dead; we will not stay.*
> *The bay-trees in our country are all wither'd*
> *And meteors fright the fixed stars of heaven;*
> *The pale-faced moon looks bloody on the earth*
> *And lean look'd prophets whisper fearful change. . . .*[17]

Hamlet's Horatio recalls a scene from *Julius Caesar,* then adds more to it:

. . . A little ere the mightiest Julius fell,
The graves stood tenantless and the sheeted dead
Did squeak and gibber in the Roman streets;
As stars with trains of fire and dews of blood,
Disasters in the sun; and the moist star
Upon whose influence Neptune's empire stands
Was sick almost to doomsday with eclipse. . . .[18]

Shakespeare's last play, *Henry VIII,* returns to the image of the falling star in *Richard II.*

During the coronation scene two gentlemen converse:

Second Gentleman:	*Those men are happy, and so are all are near her.*
	I take it, she that carries up the train
	Is that old noble lady, Duchess of Norfolk.
First Gentleman:	*It is; and all the rest are countesses.*
Second Gentleman:	*Their coronets say so. These are stars indeed;*
	And sometimes falling ones.
First Gentleman:	*No more of that.*[19]

This is a sample of the plethora of astrological references in Shakespeare's writing. Of the 135 references to stars in the canon, 47 directly invoke the power of judicial astrology.

Shakespeare expresses a particular interest in the Pleiades. In *1 Henry IV* Falstaff says:

> *Indeed, you come near me now, Hal; for we that take purses go by the moon and the seven stars, and not by Phœbus, he, "that wandering knight so fair."*[20]

In *2 Henry IV* Pistol excitedly announces to Falstaff that "we have seen the seven stars."[21] And Lear's Fool suggests:

> *Fool:* *Thy asses are gone about 'em. The reason why the seven stars are no more than seven is a pretty reason.*
> *Lear:* *Because they are not eight?*
> *Fool:* *Yes, indeed: thou wouldst make a good fool.*[22]

Does Shakespeare deliberately exploit judicial astrology just for dramatic impact, and does he not really believe in it? From Edmund's answer to his father's fear of the "late eclipses" (see chapter one) it would seem that Shakespeare is certainly aware of the argument against judicial astrology. There are other brilliant references, such as Cassius's plea to Brutus in *Julius Caesar*:

> *Men at some time are masters of their fates;*
> *The fault, dear Brutus, is not in our stars,*
> *But in ourselves, that we are underlings.*[23]

But then we have Macbeth, who thinks he can command the stars, rather than being commanded by them:

> *Stars, hide your fires,*
> *Let not light see my black and deep desires.*[24]

Shakespeare occasionally invokes the north celestial pole to show strength of character. The most famous such reference, of course, is Caesar's:

> *But I am constant as the northern star*
> *Of whose true-fix'd and resting quality*
> *There is no fellow in the firmament.*[25]

In *Lear,* Edmund claims that his "nativity was under Ursa Major."[26] Ursa Major, the Great Bear, is close in the sky to the pole star. Critic Johnstone Parr reminds us that, in judicial astrology, "the constellations north of the zodiac have their respective influences, analogous to those of the planets. . . . Ursa Major is like Mars, but the nebula under the tail resembles . . . Venus in its influence."[27] The warlike resemblance of Mars seems to fit Edmund's character, but one must be careful about Venus: the "nebula under the tail" of the bear most likely is the mass of faint stars of Coma Berenices, an entirely different constellation.

In *Romeo and Juliet,* there are so many astrological allu-

"... to hitch his wagon to a star."

—Ralph Waldo Emerson "Society and Solitude: Civilization," The Complete Works of Ralph Waldo Emerson, vol. 7 (New York: AMS Press, 1979), p. 28.

Astronomy might have been practiced in observatories like this one. England's Stonehenge is a towering and magnificent place. The ruins of what was probably an observatory and temple, Stonehenge has a long and glorious history. Beginning as a simple ditch and circle of 56 pits dug around 3100 B.C.E., the structure was expanded, rebuilt, and rebuilt again over more than two thousand years. Photo by David Levy.

sions that Shakespeare might have intended astral influence as a justification for this tragedy of apparent chance. The most famous appears in line 6 of the Prologue, where Romeo and Juliet are introduced as "star-crossed lovers." Take the name of Mercutio, who has the temper of Mercury, and the Friar's apostrophe, in which he blames Capulet's misfortunes on the stars:

> *The heavens do lour upon you for some ill;*
> *Move them no more by crossing their high will.*[28]

It is possible that Shakespeare intended the stars to have a role in other tragedies of chance, especially *King Lear,* where astral influence plays a prominent role.

In *Troilus and Cressida,* Ulysses cries out:

The heavens themselves, the planets, and this centre,
Observe degree, priority, and place,
Insisture, course, proportion, season, form,
Office, and custom, in all line of order;
And therefore is this noble planet Sol
In noble eminence enthron'd and spher'd
Amidst the other, whose med'cinable eye
Corrects the ill aspects of planets evil,
And posts, like the commandment of a king,
Sans check, to good and bad.[29]

Shakespeare was superb at weaving into his plays the sum of contemporary beliefs about the heavens. *Troilus and Cressida* was written no later than 1609, the year before Galileo's observations of Jupiter, but still a time when the vast majority of Shakespeare's audience would have been happy with a cosmological system based on "the planets and this centre" Earth. The problem, according to one critic, is that astrology had a dual role as a hybrid science with its natural and judicial branches.[30] Shakespeare's avoidance of Copernican theory, coupled with his frequent allusions to judicial astrology, do not really mean that he had no interest in contemporary science. On the contrary, these allusions indicate a fascination with the eclipses and other scientific events of Shakespeare's day and an interest in how humanity blames the stars for its own mistakes.

Chapter Three

A Changing Universe

P*adua, Italy, 1610:* Imagine a large room with a high ceiling, a desk, and a chair. Seated in the chair is a middle-aged man named Galileo Galilei, one of the most prominent scientists of his time. He has just spent two nights observing the planet Jupiter with a small telescope; this is the first time a telescope has ever been used to look at the sky. Still not quite believing his own observations, he is completing his notes on the four moons he has discovered, four moons that revolve about Jupiter and not the Earth.

Galileo had put his telescope together only a year earlier, fashioning it from two spectacle lenses placed a certain distance apart. By looking through the two glass lenses, he could magnify distant objects. With his new telescope, Galileo made one astounding find after another—craters on the Moon, the phases of Venus, and spots on the Sun.

Galileo was right to question his own observation that day. Extraordinary claims, Carl Sagan would note centuries later, require extraordinary evidence; Galileo's discovery would be one of the most astonishing claims in the history of science. In just a few nights of observation, Galileo had seen the four moons circling Jupiter. The phases of Venus also could be explained only if that planet orbited the Sun, and not the Earth. If Galileo's observations were correct, there could be but one explanation: the Earth was not the center of the universe.

Literature Responds to the New Philosophy

Although Galileo's observations brought the new ideas about Earth's position in the cosmos to the people, they were not the first indication of changing thought. Six decades earlier, Nicholas Copernicus, at the point of death, published *De Revolutionibus.* Appearing in 1543, this book set out a new way of looking at the heavens, with the Sun, not the Earth, at the center of the universe. *De Revolutionibus* was a sensation among the intellectuals of the time. It is interesting that the first political use of the word revolution appeared in the year 1600, to mean "a complete overthrow of the established government in any country or state by those who were previously subject to it."[1]

Galileo knew of Copernicus's work, and it was obvious to him that Ptolemy's system, which held that everything must revolve about the Earth, was no longer viable. He knew that Jupiter's family of moons proved that was right. However, he also knew that the time to defend Copernicus's vision had not come.

When *De Revolutionibus* was published, there was, in fact, plenty of observational evidence to support the idea that the Earth circled the Sun. The most obvious clue was that the stars rise and set some four minutes earlier each night, a small amount of time that quickly grows significant—half an hour's difference each week, and two hours each month.

The motion of the other planets through the sky is another line of evidence. Mars, Jupiter, and Saturn move in an eastward direction through the sky until they reach a certain place. Then they appear to slow down, almost stop, and next proceed westward for a period of a few months. Eventually this retrograde motion slows, then stops, and the planets resume their eastward pace. Though the reason for this would not have been apparent to people before the seventeenth century, it is evident now: in its orbit around the Sun, the Earth is overtaking the more distant, slower moving planets. The sensation is similar to that of a driver passing a slow car: as he overtakes the car, it seems to slow down and then move backward relative to the passing driver.

The great meteor showers, their meteoroids crashing through the Earth's atmosphere like spokes in a wheel, also show that the Earth is pushing its way through space. However, this esoteric fact, along with the others just mentioned, would not have been apparent to observers before the seventeenth century. Even if Copernicus's contemporaries had understood these lines of reasoning, the general population would have found little meaning in them. Thus, as a theory, the Copernican idea attracted wide attention among scientists but not among the general public or those who followed the arts.

Sir John Davies, a contemporary of Shakespeare, was an exception. He saw the new philosophy as a challenge to the magnificent cosmic dance he envisaged in his 1594 poem *Orchestra*. He gave the new ideas three lines in parentheses, and then concluded that the Ptolemaic Earth-centered system fit the order better:

> *Only the earth doth stand forever still:*
> *Her rocks remove not, nor her mountains meet;*
> *(Although some wits enrich'd with learning's skill*
> *Say heaven stands firm and that the Earth doth fleet*
> *And swiftly turneth underneath their feet)*
> *Yet, though the earth is ever steadfast seen,*
> *On her broad breast hath dancing ever been.*[2]

The important astronomical theme of *Orchestra* is that celestial bodies are dancing. In order to follow this dance we need to understand the details of the individual patterns, but how can we if the starting positions of the dance's most important partners, the Earth and the Sun, are uncertain? With the Moon orbiting about the Earth, both theories are in agreement, and Davies' poem captures their delicate orbit:

Who doth not see the measures of the Moon?
Which thirteen times she danceth every year,
And ends her pavan thirteen times as soon
As doth her brother, of whose golden hair
She borroweth part and proudly doth it wear.
Then doth she coyly turn her face aside,
That half her cheek is scarce sometimes descried.[3]

Davies' poetic description of the Moon's phase needs no scientific change after four centuries. Her dance with her brother, the Sun, involves choreographed changing positions; she faces him at full phase, then turns her face aside as the phase wanes. Would that the rest of the dance be so easily described!

Astronomical theories changed radically after Galileo's discoveries. Instead of a theoretical document, there was now a case of four moons that clearly did not revolve about the Earth. In Ptolemy's system, everything, including Jupiter's moons, should revolve around Earth. At the height of Galileo's career, this find was his greatest moment. It was also the beginning of his downfall.

The New Philosophy and Francis Bacon

During this same era, Francis Bacon was exploring a new way of looking at the natural world. He believed that ancient learning showed a shallow understanding of the world around us. This failure was not due to any lack of intellect in the minds of the ancient philosophers, but to the methods by which they built their ideas. Instead of relying on observations and experiments, these philosophers invoked instead the elegance of the argument itself. The philosopher with the most sensible reasoning, regardless of how well it related to observation, would win the day.

It is interesting to note that, for someone committed to the process of observation, Bacon enjoyed using the "old opinion" to buttress his own theoretical arguments when it was convenient for him. In his essay "Of Seditions and Troubles" Bacon writes: "When discords, and quarrels, and factions, are carried openly and audaciously, it is a sign the reverence of government is lost; for the motions of the greatest persons of a government ought to be as the motions of the planets under *'primum mobile'* according to the old opinion, which is, that every of them is carried swiftly by the highest motion, and softly in their own motion; and therefore, when great ones in their own particular motion move violently . . . it is a sign the orbs are out of frame."[4] These words were written in 1625, long after Galileo made his observations.

Although Ptolemy's system was the tried and true order of the cosmos for two thousand years, it was showing its age, and Bacon knew it. To account for the motions of the planets too complex for basic Ptolemaic theory, the astronomers of the time invented epicycles, or circles within other circles. In his essay "Of Superstition," written just two years after Galileo's discoveries, Bacon lamented that "the master of Superstition is the people, and in all Superstition wise men follow fools: . . . the schoolmen were like astronomers, which did feign eccentrics and epicycles, and such engines of orbs to save the phenomena, though they knew there were no such things. . . ."[5] Bacon used the structure of Ptolemy when it fit his own arguments, but at his most passionate he demonstrated that its arguments were hollow.

Bacon's major writing, dated 1620, was *Novum Organum*, a remarkable work for its time; in fact, critic Jean-Marie Pousseur asked whether Bacon should be credited with being the father of all modern scientific reasoning because of this writing.[6] Bacon's work is very important because it focused on the need for all philosophers and scientists to do what some, like Tycho, Copernicus, and Galileo, had already been doing. Maybe it is more accurate, in these times, to say that Bacon was the agent, not the inventor, of the modern scientific method.

Bacon wrote of two ways of inductive reasoning: the sterile "anticipation of nature," conducted by the ancient and medieval philosophers, and the superior "interpretation of nature." Bacon held that it was not up to nature to conform to a philosopher's argument, but that it was up to the philosopher to understand nature through careful observation. Thanks to *Novum Organum,* science by experiment and observation, rather than by creative reasoning and argument, had finally started along the path to respectability.

Science, however, still had quite a way to go before the rest of humanity accepted its new look. In 1611, the Catholic Church filed the first of several documents against Galileo—in secret and apparently without his knowledge. Five years later, the *Codex* against Galileo specified that any teaching that the Sun is in the center of heaven would be forbidden by the Church. Prudently, Galileo decided to wait until a more propitious time to defend the meaning of his sightings of Jupiter's moons. In 1623, Urban VIII, a modern thinker who seemed open to new ideas, was elected to the papacy. For a few years, Galileo walked and talked with the pontiff. Thinking that the time had come to speak out, late in 1629 he completed his *Dialogue Concerning the Two Chief World Systems.* One of the most important scientific works ever written, this book articulated Galileo's view of his telescope in the form of a debate. Imprudently, he created a character named Simplicio who parodied the pope's ideas.

Urban VIII was infuriated. He felt that Galileo had taken

advantage of his friendship, and that, through Simplicio, he was poking fun at the very core of Urban's papacy. The pope put the old and virtually blind astronomer at the mercy of the Holy Office of the Inquisition. Threatened by the instruments of torture, Galileo was forced to "abandon the false opinion that the sun is the center of the world," and compelled to live out his days under house arrest. Galileo would never know how important a tool his telescope would become. Almost four centuries later, a giant version of his telescope would orbit the Earth, and a spacecraft named for him would visit the moons he discovered.

On November 10, 1979, Pope John Paul II asked that the 1633 conviction be annulled, saying that Galileo "had much to suffer, we cannot hide it, at the hands of men and the agencies of the church." However, since teaching the Copernican theory had been forbidden by the Congregation of the Index in 1616, it was possible that a new trial could find him guilty again. Thus, it was suggested that the 1616 prohibition be reversed, and in 1992, it was reversed.

Galileo's discovery was a prime subject for consideration of the intellectuals of his time, and it was featured prominently in a poetic genre now known as philosophical poetry. Plato wrote of an "ancient quarrel between philosophy and poetry," and this quarrel flourished during the Renaissance. For most Renaissance writers, the poem by Lucretius, *De Rerum*

Natura (On the Nature of Things), written in the last century B.C.E., is the model didactic poem, a work whose main purpose is to describe or explain the workings of the natural world and our perception of it.

The whole genre of philosophical poetry was based, to a varying extent, on Lucretius's work, and Galileo's discovery provided fresh material for this type of poetry. John Donne's "First Anniversary" appeared only one year after Galileo's Jupiter observations, and it left no doubt that the new ideas had become prominent among serious philosophical poets who were concerned not only with changing views of the universe, but also with humankind's role in that universe:

And new philosophy calls all in doubt,
The Element of fire is quite put out;
The Sun is lost, and th' Earth, and no mans wit
Can well direct him where to looke for it.
And freely men confesse that this world's spent,
When in the Planets, and the Firmament
They seek so many new; they see that this
Is crumbled out againe to his Atomies.
Tis all in pieces, all cohaerence gone;
All just supply, and all Relation:
Prince, Subject, Father, Sonne, are things forgot,
For every man alone thinks he hath got
To be a Phoenix, and that then can bee
None of that kinde, of which he is, but hee.[7]

In 1653, John Donne wrote satirically about Copernicus in "Ignatius His Conclave," in which Ignatius enters Hell, sees all the rooms, and finds a man. "As soon as the door creaked," he notes, "I spied a certain mathematician, which till then had been busied to find, to deride, to detrude Ptolemy, and now with an erect countenance and settled pace came to the gates, . . . beat the doors, and cried, 'Are these shut against me, to whom all the Heavens were ever open; who was a soul to the Earth, and gave it motion?'"

After some thought, Lucifer answers, "what cares . . . whether the earth travel or stand still? Hath your raising up of the earth into heaven brought men to that confidence that they build new towers or threaten God again? Or do they, out of this motion of the earth, conclude that there is no Hell, or deny the punishment of sin? Do not men believe? Do they not live just as they did before? Besides, this detracts from the dignity of your learning, and derogates from your right and title of coming to this place, that those opinions of yours may very well be true. . . . Let therefore this little mathematician, dread Emperor, withdraw himself to his own company. . . ."

"Lucifer signified his assent, and Copernicus, without muttering a word, was as quiet as he thinks the sun."[8]

Donne was not alone in his worry that all coherence was gone, that the natural order was giving way to disorder. The Renaissance poet that went furthest in trying to explain the clockwork cycles of the heavens, as Copernicus saw them, was Henry More, who penned in 1647:

And as the Planets in our world (of which
the Sun's the heart and kernel) do receive
Their nightly light from Suns that do enrich
Their sable mantle with bright gemmes, and give
A goodly splendour, and sad men relieve
With their fair twinkling rayes, so our worlds Sunne
Becomes a starre elsewhere, and doth derive
Joynt light with others, cheareth all that won
In those dim duskith Orbs round other Suns that run.[9]

Although Galileo sensed that history would be on his side, John Milton was not so sure. The poet was planning an epic poem about some great subject. In September of 1638, Milton visited the scientist, who was then aged and blind. Milton was moved almost beyond words by this visit.[10] If critic Fanny Byse is correct, Galileo was saddened by the death of his daughter, a nun, and Milton would later remember Galileo's dejection in his "Il Ponseroso":

John Milton

Come, pensive Nun, devout and pure . . .
And looks commercing with the skies . . .[11]

In 1667, Milton, by now also blind, published *Paradise Lost*. In book 5 he remembers his visit with this mention of Galileo:

From hence—no cloud or, to obstruct his sight,
Star interposed, however small—he sees,
Not unconform to other shining globes,
Earth, and the Garden of God, with cedars crowned
Above all hills; as when by night the glass
Of Galileo, less assured, observes
Imagined lands and regions in the Moon;
Or pilot from amidst the Cyclades
Delos or Samos first appearing kens,
A cloudy spot.[12]

In book 7, Adam asks about the nature of the Earth, and he suspects that the Earth is not the center of the universe:

When I behold this goodly frame, this World,
Of Heaven and Earth consisting, and compute
Their magnitudes—this Earth, a spot, a grain,
An atom, with the firmament compared
And all her numbered stars, that seem to roll
Spaces incomprehensible (for such

Their distance argues and thir swift return
Diurnal) merely to officiate light
Round this opacous Earth, this punctual spot,
One day and night, in all their vast survey
Useless besides—reasoning, I oft admire,
How Nature, wise and frugal, could commit
Such disproportions, with superfluous hand
So many nobler bodies to create,
Greater so manifold, to this one use,
For aught appears, and on their orbs impose
Such restless revolution day by day
Repeated, while the sedentary Earth,
That better might with far less compass move,
Served by more noble than herself, attains
Her end without least motion, and receives,
As tribute, such a sumless journey brought
Of incorporeal speed, her warmth and light:
Speed, to describe whose swiftness Number fails.[13]

What is important about this reference to Earth's role in the universe is not which side Milton is on—*Paradise Lost* does not make that clear—but that the issue is important enough that Milton has Adam discuss it with the angel Raphael.

Seventeenth and Eighteenth Centuries

To think that science and poetry are two disciplines that are properly divorced from each other is to lose sight of what each is about and what their common goal is. In their highest forms, both are avenues of inquiry into the human condition and its relationship to the universe. Knowing what that universe is and how it is structured is fundamental to each.

By the mid-seventeenth century, telescopes were providing tremendous new insights into the solar system. In 1656, Christiaan Huygens (who a year later would invent the first pendulum clock) discovered that Saturn is circled by a ring. He announced his find in an anagram, his sentence gutted into a series of letters. When reconstructed, his discovery sentence reads:

> *Annulo cingitur, tennui, plano, nusquam coherente, ad eclipticam inclinato.*

Loosely translated, this means:

> *It is surrounded by a thin, flat, ring, nowhere touching, inclined to the ecliptic.*

Not only was Earth accompanied by a series of other worlds orbiting the Sun, but also these worlds were becoming more interesting and less esoteric. By 1700, telescopes were common enough that astronomers around the world were becoming familiar with them.

Could it have been an accident of history that two of the greatest scientific minds in history spun their magic at the same time and at the same place? For it was Edmond Halley and Isaac Newton who chatted one day about planets, comets, and gravity. Halley had examined records of comets going back two centuries or more. From some eighty-five comets that appeared between 1500 and 1700, Halley had the genius to pick out the fact that three comet apparitions were separated by about seventy-six years. The comets in question appeared in 1531, 1607, and one Halley witnessed himself, in 1682. He told Newton about this discovery, and went on to propose that the three were return visits of the same comet.

This news from Halley could not have reached Newton at a better moment, and he replied that he was working on a little project that might shed light on Halley's question. Newton's work turned out to be his *Principia Mathematica*, in which he established his laws about how worlds move about the Sun.

With its once-in-a-lifetime schedule, Halley's Comet brings generations together. The potential for grandparent and grandchild to stand together to look at Halley's Comet is repeated every seventy-six years at the comet loops around the Sun, returning, in a sense, to check on the progress of civilization. Its visit highlighted the defeat of Attila the Hun in 451 and terrified the warriors of the Norman conquest of England in 1066. We can see Halley's comet pictured on the tapestry in a church in Bayeux, France, that commemorated the conquest. It returned in 1456 and 1531, and in 1607 some bold families saw it as they were starting new lives in the new world. It was on its following trip around in 1682 that the comet was observed by Edmond Halley himself. Halley did not live to see the triumph of his prediction. The comet's appearance on Christmas night of 1758 was the final proof that it was a wandering cousin of the Earth, orbiting as a prisoner of the gravity of the Sun.

The eighteenth century brought continuing advances to our knowledge of the sky, thanks largely to constant improvements being made in telescopes. By 1781, William Herschel was well into a survey of the heavens. On March 17 of that year, Herschel noticed a

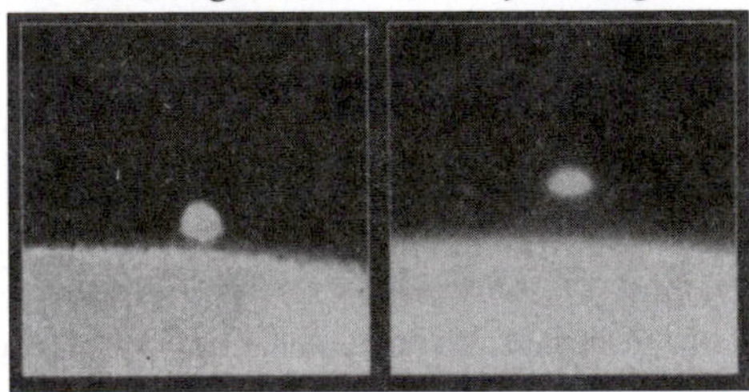

Not till the fire is dying in the grate,
Look we for any kinship with the stars.
—George Meredith (1828–1909),
"Modern Love," The Oxford Dictionary of Quotations
(London: Oxford University Press, 1955), p. 336.

On July 18, 1994, fragment G collided with Jupiter at a speed of 60 kilometers per second. NASA's Hubble Space Telescope recorded this magnificent view of the resulting plume of material that rose some 3,000 kilometers into Jupiter's atmosphere after the collision.

strange object in the constellation of Gemini, an object that turned out to be the first planet discovered in historic times. After the discovery of Herschel's planet, later named Uranus, telescopes became increasingly popular in England, which led to innovations in the instruments. George Adams and his son made a series of beautiful reflectors, with metal mirrors mounted in brass tubes. Jesse Ramsden and John Dolland crafted small refractors, with lenses that focused the light at an eyepiece at the opposite end of the tube in the style of Galileo's telescope. Standing in line to look through one of these telescopes became a popular evening activity, and, depending on the quality of the telescope and the skill of the person using it, the views ranged from disappointing to superb. In 1806 William Wordsworth commented on the crowd trying to look through a telescope in his poem "Star Gazers":

> *Whatever be the cause, tis sure that they who pry and pore*
> *Seem to meet with little gain, seem less happy than before:*
> *One after one they take their turn, nor have I one espied*
> *That doth not slackly go away, as if dissatisfied.*[14]

It has been my most serious goal, when showing first-time viewers through a telescope, that they do not leave dissatisfied. The solution now, as in 1806, is in the choice of celestial object to view. A star or galaxy would not have been too impressive as seen through the small eyepieces of Wordsworth's time. However, the Moon should have looked magnificent, then as now. For young children, I sometimes stand at the telescope so that I can see the beam of moonlight actually enter the child's eye. In this manner I make sure that the first look is so successful that even Wordsworth would have been satisfied.

As the nineteenth century went on, humankind's understanding of the heavens, especially the solar system, continued to leap forward. We began to see the Earth as one planet among seven very different worlds. In 1846, John Couch Adams, of England, and Urbain Leverrier, of France, independently completed the mathematics that led to the discovery of an eighth world called Neptune. At the same time, Charles Darwin was studying life forms on the Galapagos Islands off the coast of Ecuador, and developing his seminal work, *The Origin of Species*. Once again our ideas of a neighborhood of worlds, and of the development of life on one of these worlds, were undergoing significant change. As we shall see in the next chapter, England's great poet, Alfred Lord Tennyson, captured this golden moment in civilization when he wrote these lines in 1850:

Ring out, wild bells, to the wild sky,
The flying cloud, the frosty light:
The year is dying in the night;
Ring out, wild bells, and let him die.

Ring out the old, ring in the new,
Ring, happy bells, across the snow:
The year is going, let him go;
Ring out the false, ring in the true.[15]

Chapter Four
Ring Out, Wild Bells

I*t would have* been hard to predict that Alfred Tennyson, the son of a rector in a town on England's East coast, would become one of science's greatest expositors—a man who captured the essence of the major scientific achievements of the nineteenth century, and then returned them to the world through the words and rhymes of his poetry. During his years at Trinity College, Cambridge, Tennyson joined an intellectual group called the Apostles, and he thoroughly enjoyed the conversations he had with his friends there. But when his closest friend in that group, Arthur Henry Hallam, died unexpectedly in 1833, the young poet, stunned, became depressed and stopped publishing for almost a decade. He did not give up, however: when his writing returned, it was stronger and more mature. In 1837, he read Lyell's famous *Principles of Geology*, in which the broad outlines of the evolution of the Earth were first proposed—principles that would soon find a home in Tennyson's poems. By the 1840s, Tennyson let his lively interest in the night sky permeate his writing, as we see in his 1847 poem "The Princess":

> *Now lies the Earth all Danae to the stars,*
> *And all thy heart lies open unto me.*
> *Now slides the silent meteor on, and leaves*
> *A shining furrow, as thy thoughts in me.*[1]

During these years Tennyson was slowly building his masterpiece. In 1850, at the height of his career, Tennyson published *In Memoriam*. This was the same year he was married and was named poet laureate. *In Memoriam* was far more than an elegy to Hallam. One of the truly great poems of English Literature, *In Memoriam* begins as a memorial but ends as a celebration of mid-nineteenth-century life and the unprecedented scientific advances of that time. As critic Milton Millhauser has suggested, the poem's most profound achievement was to domesticate science, to bring it to the level of a wide readership, and to attach to it a moral and religious interpretation. It represents the highwater mark for Tennyson's enthusiasm for introducing scientific concepts into poetry, both in the amount of scientific information presented and in its poetic effectiveness.[2] Tennyson used the facts

Alfred, Lord Tennyson

of science to enrich the language of poetry, as in the following example:[3]

Who loves not Knowledge? Who shall rail
Against her beauty? May she mix
With men and prosper! Who shall fix
Her pillars? Let her work prevail.[4]

Who indeed should set the pillars of Hercules, the limits of knowledge? Six cantos later, Tennyson suggests an answer:

Let Science prove we are, and then,
What matters Science unto men,
At least to me. I would not stay.[5]

To understand the powerful role of science, particularly astronomy and evolution, in this poem, it helps to know that *In Memoriam* grew out of a collection of elegies written as the expression of agony at the loss of a friend. It was not assembled into the masterwork we know until Tennyson realized he had written so many elegies. "It is the story of a soul stunned by a tremendous loss," notes critic J. W. Pearce, "and struggling to find the meaning of a universe in which such losses can occur."[6] How common a sentiment this is! We have all have suffered tragic losses, and when we do we tend to question the natural order that allows such things to happen. (Would that the rest of us had Tennyson's skill to deal with loss in such a profound way.) *In Memoriam* goes further: it reflects the strength of its author's love of careful research into the sciences and theology. "He lacked and distrusted the passion of the revolutionary," K. W. Gransden notes, "for creating and imposing exclusive ideas of truth by skipping inconvenient evidence." Tennyson insisted on submitting the ideas of the emerging scientific thought "to the test of his own feelings and his own vision."[7]

At the start of *In Memoriam*, Tennyson puts into words his grief at the loss of his friend, but at the same time he wishes to write about a universe that brings such grief. The words from his pen are an imperfect tool for this purpose, but nature also seems to him to be structured the same way, partly to give answers, and partly to raise questions.

I sometimes hold it half a sin
To put in words the grief I feel;
For words, like Nature, half reveal
And half conceal the Soul within.[8]

Wordcraft is the poet's basic tool, and Tennyson's pen is strong enough to suggest a vision of the universe. Tennyson's vision will see the Earth both as the familiar world we live in, and as a small body in space.

This round of green, this orb of flame,
Fantastic beauty; such as lurks
In some wild poet, when he works
Without a conscience or an aim.[9]

As the poem twists and turns, its words occasionally paint a picture of a natural scene that takes this "fantastic beauty" and puts it to work:

We talk'd: the stream beneath us ran,
The wine flask lying couch'd in moss,

Or cool'd within the glooming wave;
And last, returning from afar,
Before the crimson-circled star
Had fallen into her father's grave,

And brushing ankle-deep in flowers,
We heard behind the woodbine veil
The milk that bubbled in the pail,
And buzzings of the honeyed hours.[10]

The Sun has set at the end of the day, and with the end of twilight a bright planet (the crimson-circled star) is following her father, the Sun, in setting.

In Memoriam is far more than a description of nature as elegant natural beauty. The poem is also a serious attempt to look at the scientific world of Tennyson's time, and it deals with scientific issues in remarkable detail. In the following passage, Sorrow, the priestess, suggests a pattern for the universe:

O Sorrow, cruel fellowship,
O Priestess in the vaults of Death,
O sweet and bitter in a breath,
What whispers from thy lying lip?

"The stars," she whispers, "blindly run;
A web is woven across the sky;
From out waste places comes a cry,
And murmurs from the dying sun;

And all the phantom, Nature, stands—
With all the music in her tone,
A hollow echo of my own—
A hollow form with empty hands."

And shall I take a thing so blind,
Embrace her as my natural good;
Or crush her, like a vice of blood,
Upon the threshhold of the mind?[11]

Tennyson mistrusts a universe which seems to follow a random course with no order or pattern. It is the same mistrust that John Davies noted in *Orchestra* (see chapter 3), by placing the new cosmology in parentheses (see chapter 1) so that it would not interfere with the cosmic dance of *Orchestra*. The new cosmology seems a random, cold, and silent universe in which humanity's prayers are unheard:

Man, her last work, who seemed so fair,
Such splendid purpose in his eyes,
Who roll'd the psalm to wintry skies,
Who built him fanes of fruitless prayer,

Who trusted God was love indeed
And love Creation's final law—
Tho' Nature, red in truth and claw
With ravine, shriek'd against his creed—

Who lov'd, who suffer'd countless ills
Who battled for the True, the Just,
Be blown about the desert dust,
Or seal'd within the iron hills?[12]

Sixteen years after the publication of *In Memoriam*, Gerard Manley Hopkins, a poet whose work we will explore in the next chapter, probed the mysteries of the night sky and asked:

God, though to Thee our psalm we raise
No answering voice comes from the skies;
To Thee the trembling sinner prays

But no forgiving voice replies;
Our prayer seems lost in desert ways,
Our hymn in the vast silence dies.[13]

Both poets wrote from a strong spiritual sense, coupled with a profound wonder at the sky. The cry from a dying Sun invoked by Tennyson in the stanza beginning, "The stars . . . blindly run" harks back to the nebular hypothesis, the favored theory, in that time, of how the solar system began. Formulated by Pierre Simon, Marquis de Laplace, this hypothesis held that the planets are children of the Sun, formed of the outskirts of the cloud whose center became the Sun. If that theory is correct, then the Sun is slowly losing its heat and will someday die. Although that will happen, the Sun will retain its current basic state for five billion years. What I find more interesting than this idea is the one expressed in the first statement from "Sorrow" that the stars blindly run to weave a web across the sky. In one interpretation of "blindly," Tennyson suggests that objects in the sky do not obey an intelligence but the celestial mechanics worked out by Newton, whose laws of gravitation described how objects travel through space.[14] Yet to me, this interpretation does not seem plausible. "Sorrow" describes a celestial system where nothing works right, and this the blind running of the stars is meant to suggest cosmic anarchy.

References to the nebular hypothesis appear throughout *In Memoriam*, even at places where it is invoked peripherally, as with the eddies of gas involved in the formation of the planets, appearing again in canto 128:

No doubt vast eddies in the flood
Of onward time shall yet be made.[15]

Laplace published his outline of the Nebular Hypothesis a generation before Tennyson wrote these lines of *In Memoriam.* Struck by the similarity of motion and direction of our solar system, he proposed, in 1796, a theory that the Sun and planets condensed out of a cloud. As it rotated, it would threw off a series of concentric rings, each of which condensed into a planet. Although it resembles an idea that Immanuel Kant set forth in 1755, the Laplace theory probably was developed independently and with far greater detail. It was possible, some thought, that the nebulae that were being discovered across the sky were faroff examples of solar systems in formation. By the 1830s, the whole question of the nature of the nebulae was in its infancy: we knew of the objects described by Charles Messier, the famous comet discoverer who cataloged more than one hundred comets as part of his search of the night sky, and many others listed by William Herschel. It was only in the twentieth century that we have unveiled them for what they really are: clusters of stars, clouds of gas and dust, and the distant galaxies of our expanding universe.

With his theory, Laplace was surprisingly close to providing the understanding of the solar system's genesis that persists today. The solar system, astronomers now think, began as a large dark cloud in space, sitting passively for an incredibly long time. Known as a giant molecular cloud, it was more than three hundred light years across—one of the biggest objects in the galaxy. Somewhere nearby a massive star exploded, lighting the sky as a supernova and spreading carbon into nearby space and into the giant molecular cloud. With the injection of carbon, and with the possible help of some ultraviolet irradiation, the cloud slowly began to evolve into a different place, becoming more organized and starting to rotate. Its center grew hot and massive, and became the Sun. In surrounding parts of the cloud, clumps of material accreted to one another to form the planets.

Laplace's theory suggested an ordered universe, but Tennyson's "Sorrow" still saw stars blindly running. In the years before the completion of *In Memoriam,* several bright comets had "blindly run" their courses, in a sense, to weave a web across the sky. The Great Comet of 1811, which will be discussed in chapter 7, was one such comet. Eight years later, a second bright comet rounded the Sun. However, the appearance of a great comet in 1830, just three years before Tennyson first penned this group of stanzas, and another in 1831—two great comets in as many years—was highly unusual and might have added credence to Tennyson's image of objects blindly crossing the sky. Although the first of these two comets attracted some attention, when it was brightest the 1830 comet was visible only from the southern hemisphere, heading north. The second comet appeared a few months later, and was first seen over England with a bright, short tail.

Halley's comet passed over the sky of Earth in 1835, and a magnificent comet appeared in 1843. It dominated the evening sky throughout the world that March, and as it completed its hairpin encounter with the Sun, it grew a tail that stretched for seventy million miles, almost as long as the distance between the Sun and Earth.

In 1846, a fainter comet appeared, a periodic comet whose course astronomers thought they understood well. However, as it moved through the sky it split into two pieces. At the time, a comet splitting into two was an unknown event, difficult to explain, one which could have inspired a thought about objects blindly running about the sky. (Comets are now seen to split apart with some frequency, and these events are dramatic. By far the most spectacular example of a comet's breaking up occurred in July, 1992, when, unseen by anyone on Earth, Comet Shoemaker-Levy 9 broke apart as a result of an intense encounter with Jupiter. When the comet was discovered in March, 1993, its twenty-one pieces were described as a string of pearls. The comet fragments collided with Jupiter in July, 1994. See chapter 7.)

Other major events were also taking place in astronomy at the time. In 1846, two astronomers, John Couch Adams from England and Urbain Leverrier from France, independently calculated where the eighth planet should lie. Finally, Johann Galle and Heinrich d'Arrest in Germany turned their telescope to the proper location and discovered the planet. Four years later, Tennyson alluded to the event, with "arms" referring to telescopes:

A time to sicken and to swoon,
When Science reaches forth her arms
To feel from world to world, and charms
Her secret from the latest moon?[16]

Besides Neptune, which was hardly a moon, some notable new moons were discovered during the middle years of the nineteenth century. William Lassell discovered Triton, orbiting Neptune, in 1846, the year of Neptune's discovery. Hyperion, the seventh of Saturn's moons, was detected by Harvard's George Bond in 1848, two years before the publication of *In Memoriam*. Ariel and Umbriel were found orbiting Uranus in 1851, by William Lassell. Also, between 1801 and 1849 the first ten asteroids were detected.[17] This was the new science, a litany of observations and ideas that were changing the way we look at ourselves, like the sunspots which wander like isles of night across the blinding Sun:

And was the day of my delight
As pure and perfect as I say?
The very source and fount of day
Is dash'd with wandering isles of night.[18]

The process of scientific discovery is a slow and delicate one. Although Galileo first detected sunspots in 1610, observers watched the Sun for another two hundred years before Heinrich Schwabe, an amateur astronomer from Dessau, Germany, bought a small refractor telescope in 1826. With it he began to observe the Sun in hopes of finding the elusive planet Vulcan crossing in front of the solar surface. The planet did not exist, but in 1843, Schwabe announced the existence of a cycle of sunspot activity lasting about a decade. In the story of the discovery of Neptune, the calculations of both Adams and Leverrier were not taken seriously at first. George Airy, England's astronomer royal, simply did not believe Adams, and LeVerrier had to find astronomers on his own—Johanne Galle and Heinrich d'Arrest—who were willing to conduct the search that led to the discovery. It is the beautiful process of discovery and learning that Tennyson refers to in canto 23:

When each by turns was guide to each,
And Fancy light from Fancy caught,
And Thought leapt out to wed with Thought
Ere Thought could wed itself with Speech.[19]

We also know that wherever we are, our scientific understanding is just at a beginning. Even more of a beginning is how we, on this small world, relate to the grand scheme. In one of *In Memoriam*'s most exquisite passages, Tennyson looks upon humankind as babes wandering about in the woods, insisting that we do not yet know how to think of ourselves in the universe:

O, yet we trust that somehow good
Will be the final goal of ill,
To pangs of nature, sins of will,
Defects of doubt, and taints of blood;

That nothing walks with aimless feet;
That not one life shall be destroy'd,
Or cast as rubbish to the void,
When God hath made the pile complete;

That not a worm is cloven in vain,
That not a moth with vain desire
Is shrivell'd in a fruitless fire,
Or but subserves another's gain.

Behold, we know not anything;
I can but trust that good shall fall
At last—far off—at last, to all,
And every winter change to spring.

So runs my dream, but what am I?
An infant crying in the night;
An infant crying for the light,
And with no language but a cry.[20]

This powerful thought reflects the emotions of some of us who have spent years gazing at the sky, or studying some other aspect of nature. On a dark night the sky does seem peaceful, even benevolent. It is hard to avoid connecting a starry night with the thought that all is right with the universe, and that good shall ultimately fall, and that every winter shall indeed turn to spring. But then comes our "reality check": nature is neutral, and we cannot begin to understand the meaning, if there is one, of a dark and lovely night. We are literally still infants crying in this night, in a universe so complex and vast that we can't even put the question into proper words. In that sense, Tennyson's prescient lines seek at least to frame the state of our ability to ask the question.

Canto 118 gives poetic voice to the biological breakthrough that was taking place at the very same period that Tennyson was composing *In Memoriam*. Charles Darwin's studies on the Galapagos Islands led to *The Origin of Species*, his great contribution to our understanding of our heritage and ancestry. Darwin's work led to the idea that life forms evolve from earlier forms, and that the Earth itself has evolved through time:

Contemplate all this work of Time,
The giant laboring in his youth;
Nor dream of human love and truth,
As dying Nature's earth and lime;

But trust that those we call the dead
Are breathers of an ampler day
For ever nobler ends. They say,
The solid earth whereon we tread

In tracts of fluent heat began,
And grew to seeming-random forms,
The seeming prey of cyclic storms,
Till at the last arose the man;

Who throve and branch'd from clime to clime,
The herald of a higher race,
And of himself in higher place,
If so he type this work of time

Within himself, from more to more;
Or, crown'd with attributes of woe
Like glories, move his course, and show
That life is not as idle ore,

But iron dug from central gloom,
And heated hot with burning fears,

And dipt in baths of hissing tears,
And batter'd with the shocks of doom

To shape and use. Arise and fly
The reeling Faun, the sensual feast;
Move upward, working out the beast,
And let the ape and tiger die.[21]

Tennyson's enthusiasm about the idea of evolution is tempered by his belief that some teleological force has control over the process, letting the lower life forms fall away. Tennyson didn't feel that the scientific method explains everything in nature. Urging that humanity should suppress its lower instincts, inherited from the ape and tiger, Tennyson crowns evolution with a sense of divine purpose; the evolution of the Earth is reproduced by the development of life. This is a theme that recurs in his other poetry, for example in "Maud," which appeared five years after *In Memoriam*:

So many a million of ages have gone to the making of man:
He now is first, but is he the last? is he not too base?

The man of science himself is fonder of glory, and vain,
An eye well-practised in nature, a spirit bounded and poor;
The passionate heart of the poet is whirl'd into folly and vice.
I would not marvel at either, but keep a temperate brain;

For not to desire or admire, if a man could learn it, were more
Than to walk all day like the sultan of old in a garden of spice.[22]

In staking out the grounds of his theory of evolution, Charles Darwin proposed two ideas. One was evolution itself, which says that humans moved upward, working out the beast. The second was natural selection, Darwin proposed mechanism to explain how evolution takes place. While scientists generally agree on the first, there is hot debate about the second. According to Darwin's basic theory of natural selection, species evolve gradually, without sudden change. However, evidence now exists for a series of bursts of new species. As we shall see in chapter 7, a comet or asteroid impact, or a series of damaging volcanic eruptions, could trigger such a burst.

The best known such impact took place sixty-five million years ago. In just a few minutes, an Earth teeming with many forms of life was turned into a burning wasteland. The change that occurred at the end of the Cretaceous period of Earth's history was anything but gradual. With large thunderclaps and a huge crash, a comet (or possibly an asteroid) slammed into the Earth on what is now Mexico's Yuçatan Peninsula. Kilometers-high walls of water raced out from the point of impact, and millions of tons of dust surged upward in a gigantic cloud. A crater at least 150 kilometers wide and several kilometers deep was formed in less than a minute, and excavated material from the crater rushed out with such force that it quickly circled the Earth. As the Earth's surface was blitzed with debris, temperatures rose as high as those in a broiling oven, setting off a worldwide firestorm. Within a few weeks, the whole planet was shrouded in a cloud of dust and soot. The sky was absolutely black, and for over a month there was no sunlight whatsoever, anywhere on Earth. The rain was dense with sulfuric acid. Finally, as the clouds dissipated, Earth was left with a global warming period that lasted for centuries.

By the time the episode was over, more than three quarters of the species of life on Earth had vanished. Although Tennyson was not aware of Earth's great cataclysm, he did know that at one time dinosaurs roamed the Earth, and that they were now extinct:

There rolls the deep where grew the tree.
O earth, what changes hast thou seen!
There where the long street roars hath been
The stillness of the central sea.

The hills are shadows, and they flow
From form to form, and nothing stands;
They melt like mist, the solid lands,
Like clouds they shape themselves and go.[23]

The only thing unchanging about the Earth is that it is always changing. Where once there were seas, mountains appear, and the mountains of long ago eroded down to hills and valleys. Our understanding of this continuity is not new; even the author of Ecclesiastes wrote of the majesty of Earth's water cycle, and of other cycles of change in Earth's history:

> *All the rivers run into the sea; yet the sea is not full; unto the place from whence the rivers come, thither they return again.*
>
> *The thing that hath been, it is that which shall be; and that which is done is that which shall be done: and there is no new thing under the sun.*[24]

In canto 121, Tennyson seems to be having some fun with the poetic invocation of Hesper and Phosphor, the ancient Greek names for Venus as the evening star and the morning star. Bringing epic tradition up to date means asserting that the two stars are really a single planet, Venus. As Hesper, Venus never sets long after the Sun. Tennyson knows that Phosphor can never rise during the same night; Venus is either one or the other for periods of several months at a time. But when Phosphor does rise, the "greater light" of the Sun cannot be far behind.

> *Sad Hesper o'er the buried sun*
> *And ready, thou, to die with him,*

Thou watchest all things ever dim
And dimmer, and a glory done.

. . . .

Bright Phosphor, fresher for the night,
By thee the world's great work is heard
Beginning, and the wakeful bird;
Behind thee comes the greater light.

Finally, Tennyson reveals, in the climax of this canto that the dual evening and morning star are really one:

Sweet Hesper-Phosphor, double name
For what is one, the first, the last,
Thou, like my present and my past,
Thy place is changed; thou art the same.[25]

Alfred, Lord Tennyson was a poet whose works were strengthened by repeated references to the natural world. The beauty of the night sky, as his 1842 poem "Locksley Hall" so effectively shows, was a central part of that marriage between poetry and the night sky

Many a night from yonder ivied casement, ere I went to rest,
Did I look on great Orion sloping slowly to the west.

Many a night I saw the Pleiads, rising thro' the mellow shade,
Glitter like a swarm of fire-flies tangled in a silver braid.[26]

As much as it was one of Tennyson's greatest poetic achievements, *In Memoriam* was a scientific achievement as well, for it presented the accomplishments of the science of the time to a wide readership of culturally sophisticated people who were not scientists. But to claim that *In Memoriam* was a poetic explanation of science is to limit its awesome power. The opus did not seek to explain the scientific accomplishments of the day so much as to humanize and romanticize the scientific accomplishments. Although many poets wrote of the night sky, it was Tennyson who tried to connect the romance of the sky with the accompanying scientific theories, and in doing so, he created a poetic legacy that helped to change the focus of later writing.

Is the sky a means by which we humans can see the extent of our dreams, the limit of what is possible, and even the boundless limit of the horizon of our soul? Is this the major discovery of those who share a passion for the night sky? Perhaps. We give the last word to Matthew Arnold, who suggests that the function of poetry is to ring out this thought:

Plainness and clearness without shadow of stain!
Clearness divine!
Ye heavens, whose pure dark regions have no sign

Of languor, though so calm, and, though so great,
Are yet untroubled and yet unpassionate;
Who, though so noble, share in the world's toil,
And, though so task'd, keep free from dust and soil!
I will not say that your mild deeps retain
A tinge, it may be, of their silent pain
Who have long'd deeply once, and long'd in vain—
But I will rather say that you remain
A world above man's head, to let him see
How boundless might his soul's horizons be,
How vast, yet of what clear transparency!
How it were good to abide there, and breathe free;
How fair a lot to fill
Is left to each man still![27]

Chapter Five

A Poem and a Comet[1]

August 4, 1864: A clear, dark sky. Although both Jupiter and Saturn had set, Mars was in the constellation of Aries, high in the south, its dark features visible through a telescope.[2] Auriga and Taurus were both high in the east. The Moon, having just passed its new phase, did not disturb the darkness. While vacationing in Wales that summer, a young Oxford student named Gerard Manley Hopkins might have observed this scene in those predawn hours. For all the beauty of that sky, though, Hopkins's gaze was drawn toward the second magnitude star Beta Tauri. Just west of that star shone the bright head of Tempel's comet, and its tail stretched toward another star almost as bright as the first, the nearby Iota Aurigae. The comet was moving relatively quickly, completing its quick dash past Sun and Earth within two weeks.

On September 13, 1864, less than a month after the speeding comet had faded from naked-eye view, Hopkins wrote these lines:

—I am like a slip of comet,
Scarce worth discovery, in some corner seen
Bridging the slender difference of two stars,
Come out of space, or suddenly engender'd
By heady elements, for no man knows:
But when she sights the sun she grows and sizes
And spins her skirts out, while her central star
Shakes its cocooning mists; and so she comes
To fields of light; millions of travelling rays
Pierce her; she hangs upon the flame-cased sun,
And sucks the light as full as Gideon's fleece:
But then her tether calls her; she falls off,
And as she dwindles shreds her smock of gold
Amidst the sistering planets, till she comes
To single Saturn, last and solitary;
And then goes out into the cavernous dark.
So I go out: my little sweet is done:
I have drawn heat from this contagious sun:
To not ungentle death now forth I run.[3]

Did Hopkins really see that predawn sky? We do not know. But even at that young age, he was a thorough scholar and a careful observer, and his poems reflect his keen powers of observation. Set in the Renaissance, when "single Saturn" was indeed the outermost known planet, the poem above was intended to be a speech from a play entitled *Floris in Italy*.

The speaker, Giulia, is one of a number of people attracted

to Floris, and Hopkins intended the image of the comet as Giulia's farewell soliloquy to Floris. Just as the Sun doesn't care about the planets and comets that orbit it, Floris does not recognize Giulia's worth, looking instead toward her cousin. Hopkins changed an early draft—"But when it sights the sun" to "But when she sights the sun" to strengthen the feminine metaphors of the comet image.[4] Like a stylish square dancer circling right or left around the square, Giulia "spins her skirts out" in a brave but unsuccessful attempt to win Floris, and as Giulia feels her distance from Floris growing, she falls off, shedding her smock of gold.

Hopkins was committed to the idea that every object in nature has its own individualizing quality. He called this special quality *inscape*; a comet, for example, has a pattern that is unique to it. This pattern is not just the comet's form but its inner substance and its behavior as well. The more carefully we observe a comet, the more we perceive its inscape. The response that perception invokes in us is called *instress.* Although his comet poem appeared before Hopkins developed these intriguing ideas, its lines clearly show that Hopkins saw a special relationship between the comet's appearance and its behavior. Thus, it makes sense to try to understand the comet of Hopkins's poem in relation to real comets that he might have observed, and to the rich tradition that they represent.

As we noticed in chapter 4, the 1840s were a fertile time for

PORTRAIT BY RITA MACKENZIE

Gerard Manley Hopkins

comets, and the night skies in the years preceding Hopkins's poem were even more lavish. In 1858, Donati's comet swung around the Sun with a long tail curving across a large segment of the evening sky; the famous American astronomer, George Bond, stood on the terrace of Harvard's new observatory and gazed at this ghostly marvel in the night. Only three years later, John Tebbutt, an amateur astronomer from New South Wales, sighted a comet that was moving rapidly northward. By July of 1861, this comet had a nucleus as bright as a first magnitude star, and had grown a protracted tail stretching over two-thirds of the visible sky. The comet caused a sensation in London's newspapers, which provided detailed reports and drawings of it. A year later, Comet Swift-Tuttle appeared with a rare spike directed toward the Sun, and a series of jets seemed to erupt from the comet's center. (This comet later became famous as the parent of the annual Perseid meteor stream; appearing each August, the stream consists of dust particles from this comet.) A full contemporary account of this comet appeared, curiously enough, in *Cornhill,* a literary magazine widely read by the intellectuals of the mid-nineteenth century.[5] Having enjoyed the magazine since his youth, Hopkins was one of its avid readers.

On July 4, 1864, Ernst Tempel, from Marseilles, France, discovered a comet moving northeast through the constellation of Aries; an observer named Respighi independently found the comet on the next night from Bologna, Italy. Within a month it became fairly bright; on August 8, it passed only about fifteen million kilometers from Earth, about the same distance as Comet Hyakutake in 1996. I suspect that Hopkins concentrated on Comet Tempel-Respighi, which was fresh in his memory, but that he mingled recollections of all these comets when he wrote "I am like a slip of comet." Hopkins implies that at this early stage, the comet is fragile and small, "scarce worth discovery." How true these words ring! All the comets I have discovered were faint when I first saw them, appearing as faint fuzzy patches of light, some barely brighter than the background of sky behind them.

The more I studied the poem's third line "Bridging the slender difference of two stars," the more I thought that Hopkins had a particular comet in mind, and most likely a comet he had personally observed. At first I believed the description to apply to Donati's comet, for when it was first discovered it was not far from the faint star Epsilon Leonis, moving slowly toward nearby 13 Leonis. But the line implies a comet stretching between two stars, not one appearing to move from one star to a neighboring one. Moreover, Epsilon Leonis is more than sixteen times brighter (about three magnitudes) than 13

Leonis. I could find no evidence that this early path of Donati's comet reached England's popular press of the time, and Hopkins might not have known of its early journey.

At first, Comet Tempel-Respighi, since it passed from view so quickly, seemed to be an unlikely prospect for Hopkins's thought. It was an intrinsically small comet that became conspicuous only because it passed close to Earth. (When larger Comet Hyakutake passed as close in 1996, it was quite bright and sported a tail that stretched halfway across the sky.) The 1864 comet was brighter than second magnitude for only two days, during which time it matched the sky's brighter stars. Yet, while I was searching through newspaper microfilm archives at Queen's University's Douglas library, I found this intriguing article in the *London Times*, the paper's first announcement of the comet:

> The comet first observed on the 5th inst [July] is now distinctly visible to the naked eye in the constellation Taurus and will become each night a more conspicuous object, its approach to the earth being very rapid. . . .
>
> On Monday night it will be situate [*sic*] about five degrees to the left of the Pleiades, passing thence between the stars Iota in Auriga and Beta in Taurus, towards Theta in Gemini, near which it will probably be observed on the morning of August 7. The intensity of light towards the end of the week is from 15 to 20 times greater than during our observations this morning, so that it might be expected to

> attain the brightness of stars between the first and second magnitude.[6]

It often happens that a discovery deep in the seeming abyss of a library is as exciting as a find in the depths of space. Finding this small letter in the *Times*, with its reference to the two stars, was as thrilling for me as detecting a new comet. When I mentioned my work to Joseph Ashbrook, then editor of *Sky and Telescope* magazine, he prepared an ephemeris that confirmed the *Times*' announcement that, shortly after August 1 (the "Monday night" in the extract), the comet passed between Iota Aurigae and Beta Tauri. Even though the stars' Greek letter designations indicated different brightnesses, these two particular stars had similar magnitudes. If the comet's tail followed cometary theory and pointed away from the Sun, then Hopkins might well have seen the comet and its tail "bridge the slender difference" of the two stars.

The next two lines of the poem suggest a viewpoint that dates to the Renaissance: The comet could have "come out of space" or have been "suddenly engender'd by heady elements" in the Earth's atmosphere. The play of which this poem was a part was set in that early time, and the idea that "no man knows" whether comets come from space would have been true then. By the Victorian age, it was common public knowledge, as seen in the many news stories that appeared at the time of the Great Comet of 1861, that comets were not formed from Earth's atmosphere. This major step forward was by Tycho Brahe, who demonstrated that a comet was more distant than the Moon, and later by Edmond Halley, as we have seen in chapter 3. An article in the *London Review*, appearing some four years before the poem, pointed out that the idea of comets coming from the air was proposed originally by Aristotle, and that the idea was accepted well into the Renaissance:

> Aristotle maintained that "comets" were nothing more than "meteors generated in the upper regions of the atmosphere." Seneca conceived they were real stars, but that their appearance was indicative of important changes in the affairs of mankind. "For six months," he says in his book *Naturalium Questionum*, "was this comet to be seen by us, in the happy beginning of the reign of Nero. . . ."
>
> The comet, in the estimation of the ancients, was beyond all other things in their regard, *a political star*. It was "the star of kings, emperors, and rulers; it indicated a destiny to them, and through them was supposed to affect the condition of the people over whom they presided as sovereigns."[7]

This was a time when comets were known for what they meant rather than for what they were. Lucius Annaeus Seneca had a highly personal stake in trying to attribute comets to the good graces of his patron Nero: he was Nero's tutor when, at age seventeen, Nero became emperor of Rome. During the first

He his fabric of the Heavens
Hath left to their disputes—perhaps to move
His laughter at their quaint opinions wide
Hereafter, when they come to model heaven
And calculate the stars; how they will wield
The mighty frame; how build, unbuild, contrive
To save appearances; how gird the sphere
With centric and eccentric scribbled o'er,
Cycle and epicycle, orb in orb . . .

—John Milton, *Paradise Lost* (1674), in *Paradise Lost and Selected Poetry and Prose*, ed. Northrop Frye (New York: Holt, Reinhart and Winston, 1951), p. 180, bk. 8 lines 76–84.

The large spiral galaxy Messier 88. Photo by Steve Larson and David Levy.

years of Nero's reign, Seneca had considerable power in the government, but when Nero murdered his mother, Agrippina, in the year 59, he coerced Seneca into excusing this heinous act. For the sake of his own survival, Seneca was desperate to interpret the summer comet of the following year as a good omen for Rome's ruler. Seneca spent the last years of his life with his friend Burrus, trying to limit the excesses of Nero's madness. In the year 63, Burrus died and Seneca tried to resign from politics, but Nero, feigning respect for the scientist-philosopher, refused his resignation.

Seneca knew his end under Nero's tyranny was near, and he even abandoned his wealth to live in poverty in hopes of forestalling Nero's wrath. It didn't work. In 65 C.E., Nero accused Seneca of participating in a plot against him and ordered the scholar to "prepare for death." According to custom, this gave Seneca his choice of demise. Seneca elected to cut his wrist and then let himself bleed to death, and in this contemptible fashion the life of one of cometary astronomy's greatest scholars ended.

Centuries later, long after comets were known to travel in orbits about the Sun, politics claimed the life of another cometary scientist and threatened the career of Charles Messier, the first comet hunter. Having discovered several comets, the French astronomer was famous as a gentleman scientist: Louis XV called him the ferret of comets, and he was supported by a pension from his friend, President Jean Baptiste de Saron, of the Paris parliament. De Saron was a man versed in comet orbit calculation, as well as in politics. But with the onset of the French Revolution, Messier was forced to leave the observatory in Paris. In the evening of September 27, 1793, Messier found a comet in Ophiuchus. As he had done so many times before, he informed his friend de Saron, who attempted to calculate an orbit using the positions Messier supplied. The comet was visible only briefly before it sank into the evening twilight.

However, by this time, de Saron was no longer president of the Paris Parliament. Accused as an enemy of reform, he was in prison awaiting execution by "Madame Guillotine," that creature of the Revolution. It is hard to imagine how de Saron could have cared about comet orbits when he was about to forfeit his head, but he did manage to calculate, from his prison cell, an orbit for Messier's comet. If de Saron's orbit was correct, the comet would move closer to the Sun, then swing away and reappear in the morning sky. On December 29, Messier searched the eastern sky and found his comet close to the position de Saron had predicted for it. Messier wrote of de Saron's last success and hid his note in a newspaper, which he was able to smuggle to the prisoner. On April 20, 1794, just three months before the end of Robespierre's Reign of Terror, de Saron was guillotined. Although Messier survived, his pension was gone, and the acclaimed astronomer was virtually penniless.

Seventy years after Messier's friend died, Hopkins wrote his lines about the journey of a comet that brightens as it sights the Sun. Three comets Hopkins might have seen—Donati's of 1858, Tebutt's of 1861, and Swift-Tuttle of 1862—displayed fine inner condensations that resembled a central star shaking its cocooning mists. But the "central star" of the 1861 comet was described in this way by John Challis in the *London Times* of July 3, 1861:

> The large comet which has so suddenly made its appearance in the northern sky was seen here for the first time about 10 o'clock on the night of June 30. . . . At midnight the tail reached to within a few degrees of the pole-star, and was at least 30 degrees long. The nucleus [actually the innermost

> part of the comet's head] is as bright as a star of the first magnitude. In the telescope the coma about the nucleus presents the singular appearance of four curved branches.[8]

While the "star of the first magnitude" matches the image evoked in Hopkins's line, "cocooning mists"—possibly echoing the idea of protective silky threads spun by insect larvae—are common to bright comets and were especially brilliant in the comet heads of 1858, 1861, and 1862. Since the 1864 comet was a smaller object, it had a less brilliant head.

"Shakes its cocooning mists" is a splendid image that aptly depicts the behavior of the comets of 1858 and 1861. Jets of dusty material from the heads of these comets streamed into their tails in periodic eruptions. The comet of 1858 released several gaseous envelopes at irregular intervals, but the 1861 comet shed eleven envelopes of material at the regular rate of one outburst every two days. The comet of 1862 also shook its cocooning mists, but in an unusual way: the material was apparently ejected toward the Sun instead of the more usual direction away from it. Amédeé Guillemin published this report:

> On August 10, 1862, M. Chacornac detected in the head of the comet the presence of a luminous *aigrette,* a brilliant sector directed towards the sun. . . . New sectors disengaged themselves successively from the nucleus, and on August 26

M. Chacornac determined that between the 10th and the 26th they had succeeded each other to the number of thirteen.[9]

When a comet crosses the plane of the Earth's orbit, sunlight scattered from dust surrounding the comet might cause the appearance of a sunward spike. Comet Arend-Roland, in 1957, and Comet Levy, in 1991, both displayed this unusual feature.

The phrase "Gideon's fleece," in the poem's line 11, suggests that a comet absorbs large amounts of light from the Sun, while the space around it is dark, just as the biblical fleece soaks up water while the surrounding ground remains dry:

And Gideon said unto God, If thou wilt save Israel by mine hand, as thou hast said,

Behold, I will put a fleece of wool in the floor; and if the dew be on the fleece only, and it be dry upon all the Earth beside, then shall I know thou wilt save Israel by mine hand, as thou hast said.[10]

Sometimes a comet rounds the Sun so quickly that its tail cannot keep up its role of pointing directly away from the Sun. As each dust particle in the tail assumes its own orbit, the comet leaves a tail of particles that curves away from the Sun, rather than shooting straight away from it. When the Comet of 1861 first appeared in the sky over England, it was so bright that it could be seen in daylight. The comet really did appear to "hang upon the flame-cased Sun." (The most recent comet to display this feature was Comet Ikeya-Seki in 1965; its head came within 500,000 kilometers of touching the surface of the Sun.) Comets also have gas tails made of ionized particles that rush out away from the Sun, powered by the pressure of solar radiation we call the solar wind.

By invoking Gideon's fleece, could Hopkins, who later became a Jesuit, have been speculating about how comets actually shine? In 1864 there was no consensus on whether a comet could soak up light and then re-emit it, or just reflect light from the Sun. In 1877, Amédeé Guillemin wrote that more spectral analysis was needed before one could establish whether comets shine by light other than that reflected by the Sun.[11] But he does include observations made by Father Secchi, who "concluded that, during the first few days, the nucleus shone by its own light, 'perhaps,' the Father observes, 'on account of the incandescent state to which the comet had been brought by its close proximity to the sun.'" Secchi was on the right track; a comet can light up as its gases absorb and then re-emit solar energy like a neon light.

After its close pass by the Earth, the 1864 comet faded. Hopkins uses "tether"—a brilliant choice of words—to describe the Sun's gravitational pull on the comet. A comet is bound to the Sun by a gravitational tether as surely as a dog is bound by its restraining tether. Unlike the dog who can parade about in a circle around a central pole, however, a comet orbits in an ellipse, or a parabola, with the Sun as one of the poles. Hopkins was undoubtedly familiar with the laws of celestial mechanics that Newton first proposed in the seventeenth century. Since the comet of Hopkins's poem was bound by this gravity tether, it faded as it departed from the vicinity of the Sun and the Earth. The phrase "falls off" suggests, however, that the comet was fast receding. And as the comet receded, it shredded, as Hopkins so eloquently put it, its smock of gold. As the Earth passed through the tail of the comet of 1861, the brilliant comet seemed gold-tinged. This color is rare in a comet, but it has been observed on the other occasion during which the Earth is known to have passed through a comet's tail, that of Halley's comet, in 1910.

The fourteenth line of Hopkins's comet poem originally read, "Between the sistering planets," suggesting an image of the fading comet wandering out into the outer solar system.[12] Hopkins substituted the word "Amidst" for "Between," as if to imagine the comet passing planet after planet as it receded from the Sun. We remember that the poem is set in the Renaissance, so the speaker could not have been aware of the 1781 discovery of Uranus and the 1846 finding of Neptune; to Giulia, the poem's speaker, Saturn would have been the farthest planet. The poem's closing lines relate the events of the comet's journey to that of a human life. Giulia is resigned to the end of a fine life and a satisfying relationship; like the comet, she has drawn heat from the Sun.

What may have motivated Hopkins to use the comet analogy in a play set in Renaissance Italy? Besides his fresh memory of the comet of 1864, two earlier items in the contemporary newspapers may have suggested it, one connecting the 1861 comet to Italy and the other invoking the Renaissance:

> Have we not seen quite recently, in 1861, when the great comet of that year appeared, how it was currently reported in Italy, and doubtless elsewhere, that the new star was a sign of the speedy return of Francis II and his restoration to the throne of the Two Sicilies; and also that it presaged the fall of the temporal power and the death of Pope Pius IX? "We ought not to be astonished at the persistence of these superstitions, which only the spread of science can annihilate for ever."[13]

The second newspaper item concerned a debate at the Academy of Sciences about the "terrifying" comet of 1556, which is said to have been a cause of the abdication of Charles V of Spain. With the possible exception of Genghis Khan, Charles was the Hapsburg Emperor who ruled perhaps more territories than anyone else. He retired of his own will around 1556, living quietly on his estate and indulging his interest in clocks, and died two years later. Charles's comet appeared in the southern sky of 1556, moving through Corvus and Virgo. In 1751, scientist Richard Dunthorne suggested that it was the same comet "that created astonishment throughout Europe" in 1264.[14] Dunthorne went on to prophesy a return in 1848; in 1857, with Charles's comet still unseen, J. Russell Hind, at that time one of England's best known astronomers, refined that forecast to August of either 1858 or 1860.

Even in Hind's day, astronomers understood that the gravitational pulls of the planets can affect the period of a comet as it travels about the Sun. The astronomer wrote in 1859:

> If a comet experienced no resistance while performing its journey round that luminary, it would make its appearance after equal intervals of time. . . . But the movements of comets are greatly disturbed by the attraction of the various planets belonging to the solar system, particularly by Jupiter and Saturn, which far exceed the rest in magnitude.[15]

Both the *Times* of July 4, 1861, and the *Illustrated London News,* two days later, printed these same words:

> The comet gave rise to an animated discussion at yesterday's sitting of the Academy of Sciences. . . . M. Babinet remarked that M. Hind's Ephemerides of Charles V's comet gave it the precise position of the present one. [This means that by calculating the future course of the comet that appeared in 1556,

> it would appear in the same place that the comet of 1861 now appears.] M. Bromine had predicted its return in 1858, and M. Hind admitted that it might return between 1856 and 1860. . . . If this were so, the present comet was the same that had been observed in 1556, and caused the abdication of Charles V. . . . M. Leverrier was not of M. Babinet's opinion. M. Hind's table showed different positions which Charles V's comet might occupy in the event of its return, and the question was so undetermined that it was no wonder to find a position in the table answering to that of the present comet. . . . [The] motion of the present comet . . . was so different from that given in the table, that the identity of the two comets could no longer be admitted."[16]

The scientists in this debate were well known. Most of the letters to the *Times* concerning the comet of 1861 came from Hind, who was director of the Observatory in Regent's Park, London, and had written a short book about the comet of 1556.[17] Leverrier made calculations which led to the discovery of Neptune in 1846.

Although "I am like a slip of comet" was Hopkins's only comet poem, his interest in comets lasted throughout his life. In 1874, he observed Coggia's comet: "The comet," he wrote in his journal, "I have seen it at bedtime in the west, with head to the ground, white, a soft, well-shaped tail, not big: I felt a certain awe and instress, a feeling of strangeness, flight (it hangs like a shuttlecock at the height, before it falls)."[18]

The lively image of a shuttlecock implies an object that should be in motion but which, at that second, is not moving. The comet, like the shuttlecock, appeared about to plunge down toward the horizon. Used in badminton, a shuttlecock is fitted with a ring of feathers that resembles the soft, wide tail of Coggia's comet. Note also how Hopkins is reacting to the comet's inscape through his feeling of "awe and instress."

Not long after this comet appearance, Hopkins wrote to his mother from North Wales about a possible comet discovery. No longer at Stonyhurst, Hopkins could not discuss his observation with an astronomer: "I have seen one three nights . . . in Cancer. It is small and pale but quite visible. If it is not a comet then it must be a nebula and then it is strange I should not have noticed it before. . . . At ten o'clock it is well visible in the northeast, not high; later it would be higher."[19]

This object turned out to be not a comet but the Beehive or Praesepe Cluster. Charles Messier himself included the Beehive as No. 44 in his famous list of objects that could be confused with new comets. Although Messier knew that M44 was not a comet, Hopkins should be forgiven for this interesting error. When the Beehive is low in the sky, it tends to hide its true identity of a cluster of several hundred distant stars, masquerading instead as a nebulous object, or even as a comet. In December, 1878, when the letter was written, the cluster would have been rising before dawn after several months of invisibility; Hopkins might have forgotten it from the year before. In a subsequent letter he admitted his mistake: "What I took for a comet (do you remember?) turned out to be a well known nebula of great size. Praesepe it is called, in Cancer."[20]

Hopkins's last written reference to comets appears in an 1883 letter to his friend and fellow-poet Coventry Patmore, who had included this passage in "Wedding Sermon":

> *To move*
> *Frantic, like comets to our bliss,*
> *Forgetting that we always miss,*
> *And so to seek and fly the sun,*
> *By turns, around which love should run.*[21]

Patmore's point seems to be that comets never get so close to the Sun that they are swallowed by it. Hopkins was confused by his poetic style. He thought that Patmore's image was "a contrast between the long elliptic orbits of comets, with the sun almost at one end, and the short ones, practically circles, of the planets, with the sun at the center.[22] Hopkins went on to suggest that "it might be clearer." It is true that the perihelia, or closest points to the Sun, of most comet orbits are far enough from the Sun that they do not collide. However, there is a group of comets, the Kreutz sungrazers, that might constitute a counterpoint to Patmore's idea. Sungrazing comets are exceptional because they whip around the Sun at less than one million miles from its fiery surface. First studied in detail as a single group with a possible common origin by Heinrich Kreutz in 1888, these comets share a similar orbit. Spacecraft like Solwind, SMM, and SOHO have detected hundreds of sungrazers, some colliding with the Sun. The grandeur of these comets depends on the time of year they round the Sun. Comet Pereyra, at perihelion in 1963, was bright but far from spectacular. But the Great Comets of 1882 and 1965 rounded the Sun not far from the September equinox, squeaking past the Sun at a distance less than 300,000 miles and putting on splendid shows. These two comets almost certainly last brushed the Sun as a single object, perhaps as the Great Comet of 1106.[23]

From his poems, letters, and journal writings, it is apparent that Hopkins's vigorous interest in observing the night sky profoundly affected his poetry. The detail of the astronomical imagery in "I am like a slip of comet" shows a conciseness and accuracy that befits a watcher of the sky. Hopkins's keen observational skill, coupled with his great imaginative power, added a special dimension to his writings and also to the astronomical tradition that continues to enrich the world of English poetry.

Chapter Six

Dark Sky from Walden Pond

Independence Day, 1845: Far from the reveling hordes in Boston, a twenty-eight-year-old maverick called Henry David Thoreau was trying to bring the theme of independence to a different level. For a true freedom of the mind, the young writer believed in simplifying his life by learning the lessons of nature. It is a familiar goal to many of us. We can picture a quiet lake on a sunny afternoon, a cottage with smoke wafting through a chimney, perhaps even someone setting up a small telescope at lakeside in anticipation of a starry night. We may even try to relate a bee hovering over a flowering plant, or the appearance of the Moon in the late afternoon sky, to our own experience, just like Gerard Manley Hopkins did through his inscape, and just like Thoreau tried to do in his writing.

Born in 1817, Henry David Thoreau lived his forty-four short years in Concord, a small community near a pond west of Boston. Educated at Harvard, Thoreau had a highly independent streak. When his friend, Ralph Waldo Emerson, bragged of how a Harvard education covers all the branches of learning, Thoreau quipped, "All the branches, yes, but none of the roots!"

Thoreau's first book, *A Week on the Concord and Merrimack Rivers,* was well received by critics but sold so poorly that the publisher delivered all the unsold copies—some 700 out of a print run of 1,000—to the writer's home. "I now have a library of nearly nine hundred volumes," he wrote, "over seven hundred of which I wrote myself."[1] It was only near his death that Thoreau learned that his writings were becoming more popular. His final illness began after he refused to cancel a lecture appearance when he was suffering from a cold. The cold worsened into a bad attack of bronchitis and finally mushroomed into acute tuberculosis, from which he died in 1862. Although many recognize Thoreau as one of the best of American writers, he actually left us only two well-known works. One of them, an essay called "Civil Disobedience," became the model for the careers of two of the leaders of our century, Mahatma Gandhi and Martin Luther King. *Walden,* which can be called a prose poem, is

Henry David Thoreau

Thoreau's other celebrated work. On the surface, Walden is about a writer's experience that began on July 4, 1845, when Thoreau went to live on the shore of Walden Pond near Concord. Even in the mid-nineteenth century, Thoreau's home was not terribly removed from civilization. Less than half a mile from the road leading to Concord, the location allowed Thoreau to walk to town several times a week. He wanted to demonstrate that by simplifying the details of life, it was possible to live happily without complex possessions. The fact that Thoreau took some seven years to turn Walden, the experiment, into *Walden*, the book, shows that the book is far more than an account of a year of the writer's life.

"I went to the woods," he writes in the profound opening of *Walden*, "because I wished to live deliberately, to front only the essential facts of life, and see if I could not learn what it had to teach, and not, when I came to die, discover that I had not lived." Thoreau lived in his cabin at Walden for a little over two years, with at least one interruption for a trip to the woods in Maine, and another in 1846, to spend a night in the Concord jail for refusing to pay a poll tax.

Although his years at Walden are what he is most famous for, Thoreau tried to live his entire life as modestly as possible, relating even the austerity of the night sky to his ideals. To be different merely for the sake of being different, however, was not Thoreau's way. The seed of Thoreau's convictions came from his close friendship with Ralph Waldo Emerson, one of America's most famous writers, whose philosophy included both a closeness with nature, and a truthfulness with oneself. Emerson's essay, titled *Self Reliance*, preached a solidarity with oneself:

> The first in time and the first in importance of the influences of the mind is that of nature. Every day, the sun; and, after sunset, Night and her stars. Ever the winds blow; ever the grass grows. Every day, men and women, conversing—beholding and beholden. The scholar is he of all men whom this spectacle most engages. He must settle its value in his mind. What is nature to him? There is never a beginning, there is never an end, to the inexplicable continuity of this web of God, but always circular power returning into itself. . . . The astronomer discovers that geometry, a pure abstraction of the human mind, is the measure of planetary motion. The chemist finds proportions and intelligible method throughout matter; and science is nothing but the finding of analogy, identity, in the most remote parts.[2]

The other key, to be true to oneself, is an idea that remains strong in the American consciousness:

> A foolish consistency is the hobgoblin of little minds, adored by little statesmen and philosophers and divines. With consistency a great soul has simply nothing to do. He may as well concern himself with his shadow on the wall.

> Speak what you think now in hard words and to-morrow speak what to-morrow thinks in hard words again, though it contradict everything you said to-day. "Ah, so you shall be sure to be misunderstood." Is it so bad then to be misunderstood? Pythagoras was misunderstood, and Socrates, and Jesus, and Luther, and Copernicus, and Galileo, and Newton, and every pure and wise spirit that ever took flesh. To be great is to be misunderstood.[3]

Thoreau and the Sky

"I just looked up at a fine twinkling star," an undated entry from Thoreau's journal from the 1840s says, "and thought that a voyager whom I know, now many days sail from this coast, might possibly be looking up at that same star with me."[4] During one week in 1994, as we shall see in chapter seven, this thought proved literally true, as virtually every telescope on Earth peered at the planet Jupiter. Thoreau's thought later appears in expanded form, as he launches into the diverse but related subjects of space and time: "We might try our lives by a thousand simple tests," he wrote at the opening of Walden, "as, for instance, that the same sun that ripens my beans illumines at once a system of earths like ours." Out into space but not into time: "What distant and different beings in the various mansions of the universe are contemplating the same one at the same moment! . . . Could a greater miracle take place than for us to look through each other's eyes for an instant?" And in the very next sentence, he travels first into time but not space, and next into both space and time: "We should live in all the ages of the world for an hour; ay, in all the worlds of the ages. I know of no reading of another's experience so startling and informing as this would be."[5]

Although Thoreau's published essays and books are his most famous pieces, some of his best writing appeared in his journals, which he kept for many years. While writing in a book requires drafting and careful thought, a journal allows an unfettered, spontaneous look at life, and in Thoreau's hand the journals show some of his most interesting thoughts. He would write in his cabin, or while walking. When he forgot his paper he would write on birch bark. I can't remember the number of times that humid nights have made the electric wires sizzle and almost cry out in my own, late twentieth-century nights. More than a century ago, Thoreau noted the same thing, as he wrote on September 12, 1851: "I heard the telegraph wire vibrating like an aeolian harp." On the twilight evening of June 15, 1852, Thoreau wrote: "It is candle-light. The fishes leap. The meadows sparkle with the coppery light of fireflies. The evening star, multiplied by undulating water, is like bright sparks of fire continually ascending." This evening star, subject of so many great poems, was in fact Venus that night. Earth's sister planet was in Cancer, shining at minus 4.5 magnitude, virtually as bright as it ever gets.[6] But Thoreau does not limit his description to its effect in the sky. The planet's reflection in the gently rolling waters of Walden pond offers a dramatic addition to what the sky already shows. And as the pond quiets down during the night, it reflects the images from fainter and fainter stars.

At our family cottage at Jarnac Pond in the Quebec woods, I have seen the faint fourth magnitude stars of the Little Dipper reflected in the still, mirror-like waters of the pond late at night. In any event, it is being outdoors instead of in, under the night sky instead of asleep in bed, that results in unique discoveries like planets reflecting off water: "It would be well perhaps if we were to spend more of our days and nights without any obstruction between us and the celestial bodies, if the poet did not speak so much from under a roof, or the saint dwell there so long."

Had Thoreau used a telescope that night, he would have observed Venus in its dramatic crescent phase. But it was as much a part of his purpose to inscape, as Hopkins might have said, the full impact of Venus—pond and all—without using any optical aid. *Simplify* was his clarion call: he writes proudly in Walden's opening chapter that he built his entire home for $28.12½, an enviably low price; in 1996 my fiancée, Wendee, and I spent $150,000 for our home and $6,000 more for an observatory to get as good a feel of the night sky as Thoreau had.

A feel for the sky was an important part of what Thoreau got out of Walden. On one sparkling night, for example, he found he was much closer to the skies than he was to his colleagues. He wrote:

> Both place and time were changed," and I dwelt nearer to those parts of the universe and to those eras of history which had most attracted me. Where I lived was as far off as many a region viewed nightly by astronomers. We are wont to imagine rare and delectable places in some remote and more celestial corner of the system, behind the constellation of Cassiopeia's Chair, far from noise and disturbance. I discovered that my house actually had its site in such a withdrawn, but forever new and unprofaned, part of the universe. If it were worth the while to settle in those parts near to the Pleiades or the Hyades, or to Aldebaran or Altair, then I was really there, or at an equal remoteness from the life which I had left behind.[7]

Thoreau's View of Science

Although Thoreau was quite at home with the ecosystem around his Walden home, and knew the stars and constellations well, his writings almost shunned this knowledge in favor of something more spiritual. "When I consider how, after sunset, the stars come out gradually in troops from behind the hills and woods," he wrote in his journal in 1840, "I confess that I could not have contrived a more curious and inspiring sight."[8] Curious and inspiring are two words that lead to the heart of Thoreau's concept of what the sky and its contents mean. The sky is inspiring because its vastness of time and place compels him to look inward, and, though he was as curious about the sky above as he was about the ecosystem at his Walden home, his transcendentalist philosophy dissuaded him from interpreting its physical nature. In his journals and *Walden*, Thoreau almost delights in scorning the method of science, in which objects and phenomena are observed and interpreted. An unusual thing or event, in and of itself, did not capture Thoreau's attention. For example, in the spring of 1854 he flatly refused the offer of a two-headed calf from a farmer who thought he would want to study such a freak birth from one of his cows. Thoreau's interest in nature seemed essentially a springboard to a greater understanding of self: "I am not interested in mere phenomena, though it were the explosion of a planet, only as it may have lain in the experience of a human being."[9]

> I must walk more with free senses. It is as bad to *study* stars and clouds as flowers and stones. I must let my senses wander as my thoughts, my eyes see without looking. Carlyle said that how to observe was to look, but I say that it is rather

> to see, and the more you look the less you will observe. I have the habit of attention to such excess that my senses get no rest, but suffer from a constant strain. Be not preoccupied with looking. Go not to the object; let it come to you.[10]

This concept of seeing is more of a totality of observation rather than an occasional attempt at looking. William Herschel, the famed British Astronomer Royal, felt the same way about the skill which led to his discovery of the planet Uranus: "Seeing is in some respects an art which must be learnt. To make a person see with such a power is nearly the same as if I had been asked to make him play one of Handel's fugues upon the organ. Many a night have I been practising to see, and it would be strange if one did not acquire a certain dexterity by such constant practice."[11]

Years later, Herschel expanded the concept of learning to see with one's telescope:

> It would be hard to be condemned, because I have tried to improve telescopes and practised continually to see with them. These instruments have played me so many tricks that I have at last found them out in many of their humours and have made them confess to me what they would have concealed, if I had not with such perseverance and patience courted them. I have tortured them with powers, flattered them with attendance to find out the critical moments

> when they would act, tried them with specula of short or long focus, a large aperture or a narrow one; it would be hard if they had not been kind to me at last.[12]

How does this thought differ from Thoreau's sense of what defines accurate seeing? "What I need is not to look at all," he concludes, "but a true sauntering of the eye."[13] Although Thoreau was never a scientist in the literal sense, and though he even had a disdain for the narrow research areas of some of the scientists of his time, he understood the great power of observation. This power can be aided by a telescope, as it was in Herschel's case, but it does not *need* a telescope. For Thoreau, the idea is part of a philosophy borne from the transcendentalist movement in American literature, of which he, his friend Ralph Waldo Emerson, and Nathaniel Hawthorne were lynchpins. Transcendentalism was a movement that did not flourish long—it was strongest only during the 1830s and 1840s—and it emphasized intuition, rather than knowledge, as a way of learning.

"Color, which is a poet's wealth," Thoreau emphasized in the fall of 1852, "is so expensive that most take to mere outline or pencil sketches and become men of science."[14] A week later he added, "It is impossible for the same person to see things from the poet's point of view and that of the man of science. The poet's second love may be science, not his first—when use has worn off the bloom."[15] It would seem that Thoreau's vision is completely at odds with that of this book, but it is not. In its strictest sense, the scientific method, as we discussed in chapter one, does not allow for intuition, dreaming, or excitement. But in reality science is practiced by people with imagination, humor, and varying ways of looking at nature, and there are countless cases of scientific discoveries coming out of other than official channels of research. A Japanese amateur astronomer equipped with a large pair of binoculars, for example, discovered the great comet of 1996 in time to allow professional astronomers to plan observing the rare spectacle. This was true even in Thoreau's time: the hexagonal ring-shaped chemical structure of the benzene ring, for example, came to the mind of discoverer Friedrich Kekulé in 1865 out of a dream.

I do not believe that Thoreau was showing disdain for science, but rather for what he sees as the confined way in which science was being practiced. Just as I am, in this book, trying to see the poetry in science, Thoreau dared to relate the events and objects of natural science, such as his view of Venus reflected in Walden's waters, to his own world view. Science is much more complicated now than it was in Thoreau's time, and maybe that is a good reason for us to see it in a wider sense than did Thoreau a century and a half ago.

That scientists do not have a monopoly on science is hardly a new idea. A child's first view of the heavens under a dark sky inspires a sense of wonder and awe, and the child's first longing is not for facts and answers but for a relationship: What does that sky have to do with me? Where do I fit in? Even for many scientists, this wonder never leaves, and in poetry the wonder appears again and again. Fifty years after Thoreau died, Walt Whitman wrote of the great divide between scientific theory and simple wonder:

When I heard the learn'd astronomer,
When the proofs, the figures, were ranged in columns before me,
When I was shown the charts and diagrams, to add, divide,
and measure them,
When I sitting heard the astronomer lecture with much
applause in the lecture-room,
How soon unaccountable I became tired and sick,
Till rising and gliding out I wander'd off by myself,
In the mystical moist night-air, and from time to time,
Look'd up in perfect silence at the stars.[16]

Magical thoughts such as these will not lead to an increased awareness of the ages of the stars, the nature of the Whirlpool Galaxy, or the thickness of Saturn's rings. But one doesn't need to know these things in order to sense the magic, and, beyond a very simple level, factual knowledge is not needed to appreciate the silence of the stars of Walt Whitman, nor is it needed to relate to Thoreau's transcendentalist aim. *Walden* is that rare literary form called a prose poem, and

Thoreau was a poet in the ancient sense of a prophet or seer. The poet, he wrote, was not necessarily a writer of rhyme, but was instead our interpreter of nature, able to see a beauty in nature, a force which a scientist might miss. The poet's sky was not an encyclopedia but a tabernacle.

In a sense, Thoreau's wisdom can help solve some modern-day scientific controversies. One recent debate, for example, concerns the status of Pluto, that cold and distant world. The question is, should we call Pluto a planet or an asteroid? Nature does not distinguish between minor planets and major planets, just as it doesn't differentiate between a hill and a mountain. In his journal, Thoreau agreed that we need to be as unbiased as nature is: "Not a single scientific term of distinction is distinct to the purpose, for you would fain perceive something, and you must approach the object totally unprejudiced. You must be aware that *no thing* is what you have taken it to be."[17]

I became an astronomer not to learn the facts about the sky but to feel its majesty. Perhaps skywatchers feel a part of the infinite cosmic drama most when we're with our telescopes, and our friends, at a star party. It's late, it's dark, it's clear, the Milky Way arches overhead, and we know we're where we want to be. Although it consists of many parts, the night sky as a whole is both familiar and ever-changing. In the fall of 1852 Thoreau wrote:

> After whatever revolutions in my moods and experiences, when I come forth at evening, as if from years of confinement to the house, I see the few stars which make the constellation of the Lesser Bear in the same relative position—the everlasting geometry of the stars. How incredible to be described are these bright points which appear in the blue sky as darkness increases, said to be other worlds, like the berries on the hills when the summer is ripe! . . . Far in this ethereal sea lie the Hesperian isles, unseen by day, but when the darkness comes their fires are seen from this shore, as Columbus saw the fires of San Salvador."[18]

The Hesperian Isles beckon all of us. According to one interpretation, they were the daughters of Hesperus, the most splendid star in the sky. We know from Tennyson (see chapter 4) that Hesper is the name given to Venus when that planet is in the evening sky. The daughters lived in a garden at the extreme western end of the world, and guarded the celestial flock of night. The daughters defend their secrets well; no matter how much we learn, it is not possible to know everything about the sky. The islands are magnificent: Milton describes them as "a heaven on earth,"[19] and he added:

> *Groves whose rich trees wept odorous gums and balm,*
> *Others whose fruit burnished with golden rind*
> *Hung amiable, Hesperian fables true,*
> *If true, here only.*[20]

. . . Now glow'd the firmament
With living sapphires: Hesperus, that led
The starry host, rode brightest, till the Moon,
Rising in clouded majesty, at length
Apparent queen, unveil'd her peerless light,
And o'er the dark her silver mantle threw.[21]

These are powerful islands, uniting princes, poets, and stargazers who search for something better, something transcendent, out of life. Thoreau invokes the isles of Hesper in much the same grand style of other writers of the time, but it is Tennyson who makes it personal in a poem that speaks powerfully to all of us who wish to find their dreams reflected in the western sky:

The lights begin to twinkle from the rocks;
The long day wanes; the slow moon climbs; the deep
Moans round with many voices. Come, my friends,
'Tis not too late to seek a newer world.
Push off, and sitting well in order smite
The sounding furrows; for my purpose holds
To sail beyond the sunset, and the baths
Of all the western stars, until I die.
It may be that the gulfs will wash us down;
It may be we shall touch the Happy Isles,
And see the great Achilles, whom we knew.
Tho' much is taken, much abides; and tho'

We are not now that strength which in old days
Moved earth and heaven; that which we are, we are;
One equal temper of heroic hearts,
Made weak by time and fate, but strong in will
To strive, to seek, to find, and not to yield.[22]

For a skywatcher, to visit the Hesperian Isles is to understand the precious beauty of the night. Like the idyllic lakeside cottage we dreamt of at the start of this chapter, the isles herald a night of stars, galaxies, and serenity. And if we choose to augment that feeling with knowledge and experience, then our appreciation will increase as nights pass. "Only that day dawns to which we are awake," is Walden's closing thought. "The Sun is but a morning star."

The crescent moon and Venus. Photo by Roy Bishop.

Wake! For the Sun, who scatter'd into flight
The Stars before him from the Field of Night,
Drives Night along with them from Heav'n, and strikes
The Sultán's Turret with a Shaft of Light.

Before the phantom of False morning died,
Methought a Voice within the Tavern cried,
"When all the Temple is prepared within,
Why nods the drowsy Worshipper outside?"

—Edward FitzGerald, "The Rubáiyát of Omar Khayyám of Naishápúr" (1889), reworked from the work of Omar, a twelfth-century Persian astronomer and poet. Victorian Poetry and Poetics, ed. Walter E. Houghton and G. Robert Stange (Boston: Houghton Mifflin Co., 1968), p. 343.

Chapter Seven

A Terrible Beauty Is Born

When Science Becomes Poetry

Turning and turning in the widening gyre
The falcon cannot hear the falconer;
Things fall apart; the center cannot hold;
Mere anarchy is loosed upon the world,
The blood-dimmed tide is loosed, and everywhere
The ceremony of innocence is drowned;
The best lack all conviction, while the worst
Are full of passionate intensity.

Surely some revelation is at hand;
Surely the Second Coming is at hand.
The Second Coming! Hardly are those words out
When a vast image out of Spiritus Mundi
Troubles my sight: somewhere in sands of the desert
A shape with lion body and the head of a man,
A gaze blank and pitiless as the sun,
Is moving its slow thighs, while all about it
Reel shadows of the indignant desert birds.
The darkness drops again; but now I know

Starry Night

That twenty centuries of stony sleep
Were vexed to nightmare by a rocking cradle,
And what rough beast, its hour come round at last,
Slouches towards Bethlehem to be born?

W. B. Yeats, *The Second Coming,* 1920[1]

Over the years the dual magnets in astronomy and literature have pulled my interest, and I've tried to draw these two pursuits together. By far the best such effort did not come from me, but from nature. In 1994, twenty-one fragments of the shattered comet Shoemaker-Levy 9 collided with Jupiter in the mightiest collision ever seen in our solar system.

The story of Shoemaker-Levy 9 began some 4.5 billion years ago, when a comet slowly formed out of the dust and ice in the outer part of the infant solar system. The comet spent eons as part of a sphere of comets we call the Oort cloud. For some reason—possibly the passage of a nearby star—the comet left the distant cloud and began to travel toward the inner part of the solar system. Over several orbits it passed close to the planet Jupiter. Around 1929, the same year as another historical crash here on Earth, the comet approached Jupiter in such a way that its orbit was altered. Instead of circling the Sun directly, the comet was now moving around Jupiter as a satellite.

My part of the story began one summer afternoon in 1960, when my uncle gave me my first telescope. The night that followed was pivotal. My parents and I set up the black-tubed telescope, which we named Echo after the communications satellite, in our back yard. Having no star chart, I set Echo on the brightest star in the sky. That was no star, it turned out: the four small moons that accompanied it made me certain I was looking at Jupiter.

I did not know then, nor did anyone else, that an additional moon was orbiting Jupiter that night. It was Comet Shoemaker-Levy 9. But unlike the large moons that travel in near circular orbits in regular periods, the comet cruised along a strange road that brought it to some distance from Jupiter, then sent it rushing headlong towards the planet. By 1970, I was well along in my search for comets, a calling I knew would take up many night hours, and one which might never be successful. At the time I was an undergraduate student in Acadia University's English Department, and there I found, from the pen of a long-dead writer, the justification for the road I soon took. The writer was Thoreau, and his words seemed directed at my vision:

We are all in the gutter, but some of us are looking at the stars.

—Oscar Wilde (1856–1900), *Lady Windermere's Fan* (London: Eyre Methuen, 1964), p. 51, act 3.

Comet Levy (1990c) became a conspicuous object during the summer of 1990. Photo by David Levy.

> I learned this, at least, by my experiment: that if one advances confidently in the direction of his dreams, and endeavors to live the life he has imagined, then he will meet with a success unexpected in common hours. . . .
>
> If a man does not keep pace with his companions, perhaps it is because he hears a different drummer. Let him step to the music which he hears, however measured or far away.[2]

At virtually the same time I read Thoreau's words, Comet Shoemaker-Levy 9 was passing closer to Jupiter than it ever had before, so close that it almost broke apart from the giant planet's gravity. In 1984, after searching for nineteen years, I finally discovered my first comet, and a few years later I joined the team of Eugene and Carolyn Shoemaker. For seven nights each month, we used the eighteen-inch diameter telescope at Palomar Observatory to search for comets and asteroids. Throughout the first years we spent at this telescope, our comet, still unknown to anyone, continued to orbit Jupiter, and on July 7, 1992, it headed toward the giant planet in its closest approach yet. As the comet skimmed just seventy thousand kilometers from Jupiter's cloud tops, the planet's gravity stressed the comet some two hundred times more strongly than the Moon pulls on Earth's oceans:

It is the very error of the moon;
She comes more near the earth than she was wont,
And makes men mad.

Shakespeare, *Othello*[3]

On one side lay the ocean, and on one
Lay a great water, and the moon was full.

Alfred, Lord Tennyson, "Morte d'Arthur," 1842[4]

"And the moon was full,"
As the poet said
And I aptly quoted.
And its being full
And right overhead,
Small but strong and round,
By its tidal pull
Made all being full.

Robert Frost, "Kitty Hawk," 1962[5]

So strongly did Jupiter pull on the comet that the small body began to break apart. Within two hours, what once was a single comet was now a string of twelve comets, stretching some vast distance. No one on Earth knew of the sensational opening act of the comet's final performance.

Discovery

The year 1993 began badly for the Shoemakers and me at Palomar Mountain Observatory. We were scheduled to observe the sky for seven nights in January, in the hope that at least four would be clear, yet we got only one clear night. February was even worse; of the seven nights, we had only one clear hour. During this time we prepared several dozen pieces of film for use and stored them in a light-tight box. Unknown to us, someone opened the box during the time between our February and March observing periods, and exposed the film to light.

The March observing run began well under a sparkling clear sky, but we quickly discovered that our film was useless. However, we had stored the film so that the topmost films protected the lower ones, and we were able to limp through that first night using the partially damaged films. The second night began with a fresh supply of film, but after a few hours the sky began to cloud over. As we stood outdoors looking at the sky, I suggested that we continue observing despite the weather, since the forecast was not favorable. Gene reminded me that, since we spent some four dollars each time we put a film into the telescope, which adds up to several thousand dollars a year on film, we could not afford to waste film under less than clear conditions. It was Gene's finance-based argument that led me to suggest a solution: why not use the damaged films, of which we had perhaps a dozen left? This way, we would continue to observe and waste nothing but our own time.

So it was that on March 23, 1993, I took two photographs of a region in Virgo in a partly clouded sky, on damaged films, using a telescope almost sixty years old. Jupiter was in the telescope's field of view. Two days later, Carolyn completed scanning all the films except for the last ones, and she inserted the two Jupiter films into her stereomicroscope. By scanning both films at once, in stereo, Carolyn would see the identical fields as though they were a single photograph. However, any object moving would appear to float atop the starry background. This was an elegant way of searching for comets.

It had been so long, though, since Carolyn had found anything that whenever she would adjust her chair, or clear her throat, Gene and I would look up expectantly. We did that when, just after four o'clock on the afternoon of March 25, Carolyn suddenly tensed and sat up straight in her chair. "I don't know what this is," she announced, "but it looks like a squashed comet." Gene and I took turns looking at the strangest object we had ever seen in the stereomicroscope.

In our discovery message, we wrote: "The image is most unusual in that it appears as a dense, linear bar very close to 1 arcminute long, oriented roughly east-west. No central condensation [a thickening of the comet near its center] is observable in either of the two images. A fainter, wispy 'tail' extends north of the bar and to the west."[6]

Before the comet could be announced and named, it needed to be confirmed by an independent observation. I telephoned Jim Scotti, a close friend and fellow astronomer who was observing with a larger telescope than ours from a site four hundred miles to the east of us. Skeptical at first, Jim finally agreed to turn his telescope to the coordinates I provided. Two hours later, when I called again, Jim was virtually speechless. "I am trying to pick my jaw off the floor," he said. As he described the comet's appearance—several distinct heads, each with its own tail, and two large wings of dust—I repeated each phrase to the others. In the background during this magical time, our cassette recorder happened to be playing the exultant fourth movement of Beethoven's First Symphony.

We thought that comet, eventually named Shoemaker-Levy 9, had given us its best performance just from its appearance at discovery. We were wrong. Images taken with large telescopes revealed that the comet was now split into twenty-one fragments, stretched out, one astronomer wrote, like pearls on a string. And like the falcon of Yeats's poem turning and turning in a widening gyre, the comet was orbiting Jupiter in a path that, we soon learned, had caused it to graze the planet months before we found it hurtling away.

On March 25, 1811—coincidentally 182 years to the day

before our discovery of Shoemaker-Levy 9—Honoré Flaugergues also discovered a comet. This Great Comet quickly brightened until it could be seen without a telescope, and it remained visible to the naked eye for ten months. By December, the comet had a tail more than seventy degrees long, covering a significant part of the sky. (Its appearance was even credited with the coincidentally ultra-fine wines from that year.) In October 1816, the great English poet John Keats wrote of the thrill of discovering a new work of literature, in this case a lively translation of Homer by George Chapman. Keats stayed up all night with his friend and former schoolteacher reading Homer in Chapman's translation, then he walked home at dawn, wrote the poem, and sent it to his teacher by 10 A.M. mail that same day. Most critics agree that the poet was

Hung be the heavens with black, yield day to night!
Comets, importing change of times and states,
Brandish your crystal tresses in the sky
And with them scourge the bad revolving stars
That have consented unto Henry's death!
King Henry the Fifth, too famous to live long!
England ne'er lost a king of so much worth.
—Shakespeare, 1 Henry VI , 1.1.1–7

If Shakespeare wrote Henry VI by 1592, these opening lines might hark back to major comet appearances in 1587, 1590, and 1591; a magnificent comet appeared near the North celestial pole in 1593. By no stretch of the imagination could any of these comets have resembled this image of Comet Shoemaker-Levy 9, taken by Jim Scotti on March 30, 1993, less than a week after its discovery.

probably comparing that thrill of discovery to the feeling William Herschel had on his finding of Uranus in 1781, an event that occurred fourteen years before Keats was born. I believe that the poet had a different discovery in mind, that of the Comet of 1811, when he penned his famous sonnet "On First Looking into Chapman's Homer."

John Keats

Whether the discovery lies between the covers of a book, or on photographic films at an observatory, or in the eyepiece of a telescope, the thrill is the same:

Much have I traveled in the realms of gold,
And many goodly states and kingdoms seen;
Round many western islands have I been
Which bards in fealty to Apollo hold.
Oft of one wide expanse had I been told
That deep-brow'd Homer ruled as his demesne;
Yet did I never breathe its pure serene
Till I heard Chapman speak out loud and bold:
Then felt I like some watcher of the skies
When a new planet swims into his ken;
Or like stout Cortez when with eagle eyes
He stared at the Pacific—and all his men
Looked at each other with a wild surmise—
Silent, upon a peak in Darien.[7]

Impact

On May 22, 1993, the Shoemakers and I were again observing at the eighteen-inch when the announcement came in: Comet Shoemaker-Levy 9, it turned out, was about to fulfill Yeats's prophecy in a strange way. Like the falcon unable to hear its falconer, the comet was skating out of control. Already, things had fallen apart; the center couldn't hold as the comet was continuing to split apart. Its brightest fragment was breaking up into several pieces. The fragments, we learned that day, would move away from Jupiter until July, then start to approach Jupiter again. In July 1994, the comet would crash into Jupiter. We had fourteen months to ponder what "mere anarchy" the comet would unleash on that distant world as it slammed into Jupiter at the unheard of speed of sixty kilometers per second.

As the time to impact week shortened, astronomers prepared their telescopes for observing the event. Clearly this kind of event was so rare that it demanded maximum observing time, on as many telescopes as possible. Almost every major telescope on Earth was poised to watch Jupiter in the most massive observing campaign for a single event since Galileo first turned his telescope to the sky more than four hundred years ago. In addition, an armada of spacecraft, including the Hubble Space Telescope and the Galileo space probe, were ready to turn their sights toward Jupiter. When we arrived at the

Space Telescope Science Institute in Baltimore, Maryland, on the afternoon of July 16, 1994, we found the place overrun with reporters and broadcasters from all over the world. It was in this tense atmosphere that we awaited the results from Jupiter, where Fragment A of Comet Shoemaker-Levy 9 was hurtling into the atmosphere at more than 130,000 miles per hour. One reporter teased us about how embarrassed we would be if the show turned out to be a dud. We were as concerned as he that as big as the event was, astronomers might not be able to detect any of it from Earth. We had two reasons for this concern. One was that the comet was so small, and Jupiter so large, that no matter how big the comet's punch it would leave no trace of its fiery plunge. Our other concern was that the comet was hitting the side of Jupiter facing away from Earth, so that whatever show there was might end before the planet rotated the impact site to face Earth. Would Comet Shoemaker-Levy 9 end as T. S. Eliot wrote, in 1925, in "The Hollow Men"?

This is the way the world ends
This is the way the world ends
This is the way the world ends
Not with a bang but a whimper.[8]

Finally, late in the afternoon of July 16, the good news came briskly from Europe. Observers at a telescope in Spain had seen a major explosion on Jupiter: lasting the better part of an hour, the convulsion was as bright—in the filter they used—as one of Jupiter's moons. A few hours later, the Hubble Space Telescope sent down images of a huge fireball arcing over the planet, followed by a dark cloud a quarter the size of Earth. We were in the midst of a press conference when the ecstatic astronomers a floor below received these pictures. One of these astronomers, Heidi Hammel, broke into our press conference holding a picture in one hand and a bottle of champagne in the other, and the world watched as the audience of reporters cheered this stunning news. After the press conference, we visited the observation support system room, one floor below, where images of the new dark spot left by Fragment A were displayed, and then we drove to the U.S. Naval Observatory in Washington, fifty miles away, where we looked at Jupiter through one of the great refractor telescopes there. July 16, 1994, was an unforgettable night.

As one fragment after another struck Jupiter, impact week grew in excitement. Fragment G, a hunk of ice and dust perhaps a kilometer across, struck Jupiter on July 18, leaving a cloud larger than the Earth. It was the most conspicuous marking ever seen on another planet. By the end of impact week, Jupiter lay bombarded with these dark clouds, markings that remained visible for almost a year.

Thou too, O Comet beautiful and fierce,
Who drew the heart of this frail Universe

Towards thine own; till, wrecked in that convulsion,
Alternating attraction and repulsion,
Thine went astray and that was rent in twain;
Oh, float into our azure heaven again!

Percy Bysshe Shelley, "Epipsychidion"[9]

A Lesson in Our Heritage

Although the real lessons of this extraordinary event took a while to emerge, it was evident right away that we had experienced the most fundamental of events in our solar system. Impacts are the means by which the Earth and the other planets were originally put together from basic building blocks. However, the primordial Earth was too hot for any of the building blocks for life to assemble there: no carbon, hydrogen, oxygen, or nitrogen. These materials were still intact only among the distant outer worlds of the solar system, and without any way to get them here, life would never have started on Earth. It is possible that comets, rich in these materials, provided the delivery service for the building blocks of life, both through direct crashes with Earth and through the tiny specks of dust comets leave behind, cometary particles that slowly wafted down through the atmosphere. That comets can deposit the hydrogen and oxygen of water directly into a planetary atmosphere has now been proven, for several telescopes detected, high in Jupiter's atmosphere, deposits of water from Comet Shoemaker-Levy 9.

It is a happy coincidence that the leading scientist of impact theory, Eugene Shoemaker, was one of the discoverers of the comet that collided with Jupiter. His career began when he was mapping the early nuclear bomb craters near Las Vegas, Nevada, sites that gave him the idea that some of the great natural craters on Earth and on the Moon might be the results of impacts. In the mid-1950s, he established that the crater near Flagstaff, Arizona, was the result of the crash of an asteroid from space some fifty thousand years ago. He later established the impact nature of most of the craters we see on the Moon. Shoemaker was a strong believer in the idea that the reason life exists here on Earth is that, when comets struck the Earth billions of years ago, they delivered the building blocks of life. "We are the progeny of comets," said Shoemaker, whose dream was to see an impact taking place somewhere on Earth, and then rush out to map the new crater. He never thought that his dream would be realized so dramatically, on another world, by a comet he helped discover. It was a collision that brought home the incredible power of comet impacts to bring—and to take away—life on Earth, as Frost wrote, "Some say the world will end in fire."[10]

Aftermath

By the end of Comet Shoemaker-Levy 9's impact week, Jupiter had suffered twenty-one major explosions as the fragments struck at the average rate of one every six hours, interrupting the ebb and flow of Jupiter's atmosphere like the rebounding hail on the rocks of "Kubla Khan":

In Xanadu did Kubla Khan
A stately pleasure dome decree:
Where Alph, the sacred river, ran
Through caverns measureless to man
Down to a sunless sea. . . .
And from this chasm, with ceaseless turmoil seething,
As if this earth in fast thick pants were breathing,
A mighty fountain momently was forced:
Amid whose swift half-intermitted burst
Huge fragments vaulted like rebounding hail,
Or chaffy grain beneath the thresher's flail:
And 'mid these dancing rocks at once and ever
It flung up momently the sacred river.[11]

After that incredible week of impacts, I returned home to see the large, black impact sites through Echo, the same telescope through which my parents and I first looked at Jupiter thirty-four years earlier. The telescope was still working well,

A total eclipse of the Sun, seen at sunrise on the Atlantic Ocean on August 11, 1999, reminds us that both the Moon and Sun can cause great tidal effects on Earth. The disruption of Comet Shoemaker-Levy 9 is a direct result of the mighty tidal forces of Jupiter. Photo by Roy Bishop.

and Mother enjoyed viewing the impacts with me. Dad, however, could not, for he had died nine years earlier from Alzheimer's disease. Looking through that telescope, I saw a planet vastly different, and suddenly I missed my father very much. Even though Dad had been a cautious person, wanting his family to make secure choices rather than chancy ones, he would have been pleased that my teenage decision to begin a life-long search for comets had worked out so well.

> *Two roads diverged in a wood, and I—*
> *I took the one less traveled by,*
> *And that has made all the difference.*
>
> Robert Frost, "The Road Not Taken"[12]

Anyone who saw Jupiter in the two months following the S-L 9 impacts could see that the planet looked changed in a way I had not seen in all my years of observing Jupiter. Scholars who studied the history of Jupiter observations since Galileo first examined the planet with a telescope in 1610 agreed that with the possible exception of an impact scar observed by the observer Cassini in 1690, humans had not seen an event like this before. It was as though we were looking at a new Jupiter,

> *changed, changed utterly:*
> *A terrible beauty is born.*[13]

Epilogue

The lights of heav'n (which are the world's faire eyes)
Looke down into the World, the World to see;
And as they turne, or wander in the skies,
Survey all things that on this center bee.

And yet the lights which in my towre do shine,
Mine eyes which view all objects, night and farre;
Looke not into this little world of mine,
Nor see my face, wherein they fixed (grave accent) are.

—Sir John Davies, *Nosce Teipsum*, 1599[1]

Although Sir John Davies is posing, in his philosophical masterpiece *Nosce Teipsum*, the question of how we can look inside ourselves to determine the nature of the soul, I am quoting these two stanzas for a different purpose. When I look toward the stars, which is virtually every clear night, I sense, in friendship, that they are peering back at me. This is not a scientific statement, but I do not spend hour after hour, year after year, looking skyward, just in the name of science. There is both science and poetry to the observing I do. The scientific method of inquiry has kept me interested in the nature of the sky and it contents over the years, but the passion of poetry is what draws me out each evening and morning to savor what the night sky has to offer.

The poetic side to the sky is what this book is primarily about. I did not try to quote Shakespeare or Tennyson in any effort to prove that Shakespeare believed in the Copernican Theory, or that Tennyson believed in a specific relation between science and theology. My purpose is to show only that they were attracted to the night sky, gave some thought to it, and that the sky had an impact on their writing.

For his epic poem *Paradise Lost,* for example, John Milton wrestled with two possible explanations for how the solar system works. By 1665, when the poem was first published, the Sun-centered system would have seemed to be obvious, but Milton was writing an epic for the ages and hedged his bets. Thus, in possibility 1, the angel Raphael suggests a picture of

The Earth,
Though, in comparison of Heaven, so small,
Nor glistering, may of solid good contain
More plenty than the Sun that barren shines.[2]

But in possibility 2, Raphael poses the ideas of Copernicus and Galileo:

> *What if the Sun*
> *Be centre to the World, and other stars,*
> *By his attractive virtue and their own*
> *Incited, dance around him various rounds?*
> *Their wandering course, now high, now low, then hid.*
> *Progressive, retrograde, or standing still,*
> *In six thou seest; and what if, seventh to these,*
> *The planet Earth, so steadfast though she seem,*
> *Insensibly three different motions move?*[3]

With its philosophical quandary concerning the nature of the universe, the seventeenth century—the age of Davies, Shakespeare, Donne, and Milton—had a special interest in humanity's relationship to the cosmos. That relationship has evolved over later centuries. In the pages of this book, we have seen how artists express their visions of the sky, whether in the sentences of Thoreau, the verse of Hopkins, or the canvas of Van Gogh. As always, even though the artist is expressing a personal connection with the sky, this connection often goes well beyond the individual and represents a statement of the age. For Milton, a half century after Galileo observed moons orbiting Jupiter, not Earth, the question of whether the universe was heliocentric or geocentric was still undecided. Since this debate represented a fundamental shift in humanity's understanding of its place in the universe, this delay is not surprising.

Four centuries later, that place is still not a sure thing. At the opening of the twentieth century, Henrietta Leavitt, of Harvard, noticed something peculiar about twenty-five "Cepheid variable stars," so named after the star Delta Cephei. These stars were in that nearby galaxy called the Small Magellanic Cloud, and they change in brightness like clockwork, from brighter to fainter and back again at precise intervals. Leavitt found something amazing: the brighter the average magnitude of a star was, the longer it took that star to complete a cycle of changing brightness. She brought this discovery to her colleague Harlow Shapley.

Since all the stars in the Small Magellanic Cloud are about the same distance from us, Shapley concluded that this correlation between their periods of variation and their average magnitudes could be used as a yardstick for measuring vast distances. By comparing a star's average apparent brightness with its period of variation, Shapley could calculate its real average brightness (which is not dependent on its distance). Shapley reshaped our understanding of the position of our solar system within our galaxy. The Earth and Sun, he realized, are at the edge, not in the center, of our galaxy.

The next step came just a few years later. Before the twentieth century was a quarter complete, Edwin Hubble determined that our galaxy and its neighbors were a tiny part of an expanding universe whose groups of galaxies are racing away from one another. Humanity's evaluation of its place in the cosmos, a process that began with Copernicus and Galileo, is still continuing today. Over the years, that question has focused on different areas of astronomy.

As we have seen, Tennyson's *In Memoriam* focused on the then-new nebular hypotheses theory to explain the origin of the solar system, and its relation to Darwin's theory of evolution. At the opening of the twenty-first century, humanity's place and origins have focused on the new idea of cosmic impact. In 1980, Walter and Luis Alvarez proposed that the course of life on Earth was severely affected by the impact of an asteroid or comet from space.[4] This astonishing discovery, followed in 1991 by the finding of the crater from that crash, helped focus the attention of the scientific community on the fact that the course of life can be changed by cosmic impacts.

Observations of Halley's comet by spacecraft in 1986 added another element to this new cosmic picture: the comet was found to contain "CHON" particles, or carbon, hydrogen, oxygen, and nitrogen. When comets collide with planets, they deposit their organic materials on the worlds they hit. That happened apace when the Earth was young, and is probably how the raw materials of life were brought to Earth. The collision of Comet Shoemaker-Levy 9 with Jupiter, as we have seen, was a demonstration of the process of the origin of life itself.

So, we return to Comet Shoemaker-Levy 9 and its mighty collision with Jupiter. Much has happened in the years that followed the collision of Shoemaker-Levy 9 on Jupiter. As late as 1996, Jupiter still showed signs of increased carbon sulfide because of the impacts. Back on Earth, Carolyn Shoemaker, Wendee my wife, and I are still searching for comets using our Shoemaker-Levy double cometograph, an instrument that represents the last project that Gene Shoemaker started. Each clear, moonless morning finds me outside, with telescope, fulfilling my shift as night watchman as I search the eastern sky for new comets. Each dark night is a silent page of poetry, its rhythm aligned with those distant lights in space and time.

What new ideas will those distant lights bring in future years that will find their way on a poetic page? Some of those lights, yet unseen by human eye, are comets on their way to grace the sky of Earth with poetry and art. There is something wondrous about a lonely comet spending billions of years at the edge of the solar system, then imperceptibly changing its orbit, thanks to some gravitational tug, and starting its billion-year journey to the warmth of the Sun. Near the end of that journey, an amateur astronomer awakens on Earth, pulls back his observatory roof, removes the covers from his telescope, and begins his morning of searching the sky. In the next hour his telescope will have passed over that comet, allowing it to reveal its presence.

Discovering a comet is receiving a gift from nature. In capturing the spirit of this gift in his 1995 poem "The Comet Hunter's Call," amateur astronomer Peter Jedicke used phrases from Gerard Manley Hopkins's poem "I am like a slip of comet" to form a completely different message. Hopkins's poem can be read as a journal from a comet as it rounds the Sun in space; Jedicke takes those words to form a call from comet seekers to the comets they hope to find. Sung to the tune of "Amazing Grace," these lines turn the poetic search for a relation between humanity and the cosmos into a personal relationship between one human and one comet.

A poem, wrote Percy Bysshe Shelley, "is the very image of life expressed in its eternal truth." For Davies, who wrote for all ages in *Nosce Teipsum*, and for Milton, whose message spoke to the future with *Paradise Lost*, for all the poems discussed in these pages, that truth is from the poet's mind and from the poet's heart. The passage of time does not diminish the poem's message, but augments it, as Shelley added, "and for ever develops new and wonderful applications" of it.[5] Each poet brings to the ages his or her view of our place in the cosmos. That relation is a personal one, so that in each poem we see, as in *Nosce Teipsum*, how the "lights of Heav'n . . . looke down into the world."

Starry Night

Awake, o ye comets, deep in space,
from slumber dark and cold.
Come fly to me, discovery waits.
Come don your smocks of gold.

'Tis time, my friend, to leave Oort's cloud,
where no one knows your name.
Your central star has called you down
to fields of light and fame.

So spin your skirts across the sky
and feel the contagious sun.
I shall watch for you with patient eye,
and then tell everyone!

Awake, o ye comets, deep in space,
from slumber dark and cold.
Come fly to me, discovery waits.
Come don your smocks of gold.[6]

Notes

Preface

1. Etan Weinreich, producer, *Asteroids: Deadly Impact,* National Geographic Television, aired on NBC February 1997.

2. Leslie C. Peltier, *Starlight Nights* (New York: Random House, 1965), p. 231.

3. William Shakespeare, *Romeo and Juliet,* in *William Shakespeare: The Complete Works,* ed. Peter Alexander (London: Tudor Edition of Collins, 1951), 3.2.20–25. Except for the fourth quarto edition of this play, the original texts have the second line of the quote saying, "Give me my Romeo, and when I shall die," a statement that significantly alters the meaning of the text. The second quarto appeared in 1599 and is considered the authoritative text, but it does have some problems, including this one, that the fourth quarto attempted to correct. The text that was sent as Shakespeare's first voyage into space and to the Moon follows:

And, when he shall die,
Take him and cut him out in little stars,
And he will make the face of heaven so fine
That all the world will be in love with night,
And pay no worship to the garish sun.

4. Sarah Williams, "The Old Astronomer," in *Twilight Hours: A Legacy of Verse* (London: Strahan and Co., 1869), pp. 68–71.

Chapter One

1. Thomas Gray, "Elegy Written in a Country Churchyard," *English Romantic Poetry and Prose*, ed. Russell Noyes (New York: Oxford University Press, 1956), p. 47.

2. Stephen James O'Meara, *Deep Sky Companions: The Messier Objects* (Cambridge, England: Cambridge University Press, 1998), p. 160.

3. Nathaniel Harris, *The Masterworks of Van Gogh* (New York: Smithmark publishers, 1996), p. 209.

4. Ibid., p. 210.

5. David H. Levy, *Daily Journal* (unpublished), 22 May 1973.

6. *Sky and Telescope* 70 (August 1985): 105.

7. Ibid., 76 (October 1988): 406–408.

8. Ibid., 77 (March 1989): 237.

9. Joseph Lutz, *Sky and Telescope* 78 (July 1989): 5–6.

10. Harris, *The Masterworks of Van Gogh*, p. 8.

11. Robert Frost, "A Star in a Stone-Boat," in *Robert Frost: Collected Poems, Prose, and Plays*, ed. Richard Poirier and Mark Richardson (New York: Library of America, 1995), p. 162, lines 1–4.

12. John Donne, "Go and Catch a Falling Star," in *Poems and Prose* (New York: Everyman's Library of Alfred Knopf, 1995), p. 14, lines 1–9.

13. George Gordon, Lord Byron, "Manfred," in *Lord Byron: Selected Poems and Letters*, ed. William H. Marshall (Boston: Houghton Mifflin, 1968), p. 232, 1.1.192–201.

14. Robert Southey, "St. Antidius, the Pope, and the Devil," in *The Poetical Works of Robert Southey, with a Memoir of the Author* (Boston: Little Brown, 1863), p. 147.

15. Frost, "Acquainted with the Night," p. 234.

16. Samuel Taylor Coleridge, "The Rime of the Ancient Mariner," in *English Romantic Poetry and Prose,* ed. Russell Noyes (New York: Oxford University Press, 1956), p. 396, part 4 lines 263–66.

17. Gerard Manley Hopkins, "A fragment of anything you like," *The Poetical Works of Gerard Manley Hopkins,* ed. Norman H. Mackenzie (Oxford: Clarendon Press, 1990), p. 11.

18. Thomas Gray, "Journal of a Tour Through the English Lakes," in *The Poetry of Earth: A Collection of English Nature Writings,* ed. Edward Dudley Hume Johnson (New York: Athenaeum, 1974), pp. 51–52. See also John Milton, *Samson Agonistes,* in *Paradise Lost and Selected Poetry and Prose,* ed. Northrop Frye (New York: Holt, Rinehart and Winston, 1951), p. 378, lines 86–89. Norman MacKenzie pointed out the Gray-Milton connection to me.

19. Coleridge, "Anima Poetae," in *The Poetry of Earth,* p. 129.

20. Hopkins, *The Journals and Papers of Gerard Manley Hopkins,* ed. Humphrey House and Graham Storey (London: Oxford University Press, 1959), p. 218.

21. James Kinsley, "The Ballad of Sir Patrick Spence," in *The Oxford Book of Ballads* (Oxford: Clarendon Press, 1969), p. 312.

22. Coleridge, "Rime of the Ancient Mariner," p. 395, part 3 lines.

23. *Othello,* 2.3.177–80.

24. *King Lear,* 1.2.1–2.

25. Ibid., 1.2.101.

26. Ibid., 1.2.115.

27. John Milton, *Samson Agonistes,* in *The Norton Anthology of English Literature,* ed. Meyer Howard Abrams (New York: W.W. Norton, 1968), p. 706, line 80.

28. Milton, *Paradise Lost,* in *Paradise Lost and Selected Poetry and Prose,* pp. 21–22, bk 1 line 597.

29. Thomas Hardy, "At a Lunar Eclipse," in *The Collected Poems of Thomas Hardy* (New York: Macmillan, 1926), pp. 105–106.

30. Robert Browning, "My Star," in *Victorian Poetry and Poetics,* ed. Walter

E. Houghton and G. Robert Stange (Boston: Houghton Mifflin Co., 1968) p. 220.

31. Mary E. Lozer, "Thirsting," unpublished manuscript, 1999. Reprinted by permission.

32. John Keats, "Bright Star," in *English Romantic Poetry and Prose,* p. 1189.

33. Stephen Crane, "A Man Said to the Universe," in *The American Tradition in Literature,* vol. 2, ed. Sculley Bradley, Richmond Beatty, and E. Hudson Long, (New York: Norton, 1967), p. 944.

34. Max Ehrmann, "Desiderata," in *The Poems of Max Ehrmann,* ed. Bertha Pratt Ehrmann (Boston: Boston Crescendo, 1948), p. 83.

35. Robert Browning, "Andrea del Sarto," *in Victorian Poetry and Poetics,* p. 246.

Chapter Two

1. 1 Chronicles 21:16.

2. Amos 8:9, translation from *The Holy Scriptures According to the Masoretic Text* (1917; reprint Philadelphia: Jewish Publication Society, 1955).

3. Isaiah, 38:8.

4. Geoffrey Chaucer, *The Miller's Tale,* in *The Works of Geoffrey Chaucer,* ed. Fred Norris Robinson (Boston: Houghton Mifflin, 1957), p. 51, lines 3457–61.

5. *Julius Caesar,* 2.2.31.

6. *All's Well That Ends Well,* 1.1.178–92.

7. Marjorie Hope Nicolson, "The Discovery of Space," *Medieval and Renaissance Studies,* ed. O. B. Hardison Jr. (Chapel Hill: University of North Carolina Press, 1966)

8. Theodor Ritter von Oppolzer, *Canon of Eclipses* (New York: Dover, 1962) charts pp. 131–34, lunar eclipse table p. 368.

9. *Antony and Cleopatra,* 3.13.152–54.

10. *Hamlet,* 2.2.115–18.

11. Marvin Spevack, *A Complete and Systematic Concordance to the Works of Shakespeare* (Hildesheim: Georg Olms, 1968).

12. Wolfgang H. Clemen, *The Development of Shakespeare's Imagery* (London: Methuen and Co., 1951), p. 93.

13. *King Lear,* 1.2.133–43.

14. J. W. Draper, "Shakespeare's Star-crossed Lovers," *Review of English Studies* 15 (1939): 20.

15. W. D. Smith, "The Elizabethan Rejection of Judicial Astrology and Shakespeare's Practice," *Shakespeare Quarterly* 9 (1958): 159–76.

16. *Two Gentlemen of Verona,* 2.5.73–75.

17. *Richard II,* 2.4.7–11.

18. *Hamlet,* 1.1.114–20.

19. *Henry VIII,* 4.1.49–55.

20. *1 Henry IV,* 1.2.12–15.

21. *2 Henry IV,* 2.4.177.

22. *King Lear,* 1.5.33–36.

23. *Julius Caesar,* 1.2.139–41.

24. *Macbeth,* 1.4.50–51.

25. *Julius Caesar,* 3.1.60–62.

26. *King Lear,* 1.2.123.

27. Johnstone Parr, *Tamburlaine's Malady and other Essays on Astrology in Elizabethan Drama* (Alabama: University of Alabama Press, 1953), pp. 80–81.

28. *Romeo and Juliet,* 4.5.94–95.

29. *Troilus and Cressida,* 1.3.85–94.

30. Robert Walker, "The Celestial Plane in Shakespeare," *Shakespeare Survey* 8 (1955): 109–17.

Chapter Three

1. *Oxford English Dictionary* suggests that the first such use was in Edward Blount's 1600 translation of a work by Conestaggio about the end of the House of Portugall. See G. Conestaggio, *The historie of the uniting of the kingdom of Portugall to the crown of Castill,* trans. E. Blount (London: A. Hatfield, 1600).

2. Sir John Davies, *Orchestra, or A Poem of Dancing,* ed. Eustace M. W. Tillyard (London: Chatto and Windus, 1947), p. 27.

3. Ibid., p. 25.

4. Francis Bacon, "Of Seditions and Troubles," *Essays or Counsels Civil and Moral of Francis Bacon,* (London: Collins, n.d.), p. 130.

5. Bacon, "Of Superstition," 157.

6. Jean-Marie Pousseur, *Bacon, 1561–1626: Inventer la science* (Paris: Belin, 1988).

7. John Donne, "The First Anniversary," in *The Poems of John Donne,* ed. Sir Robert Grierson (London: Oxford University Press, 1933), lines 205–18.

8. Donne, "Ignatius His Conclave," in *Donne: Poems and Prose* (New York: Everyman's Library of Alfred Knopf, 1995), pp. 199–203.

9. Henry More, "The Argument of Democritus Platonissans, or The Infinity of Worlds," (Verse 24) *Philosophical Poems* (Cambridge: Roger Daniel, printer to the University, 1647; reprint Scolar Press, 1969), p. 197.

10. "During his visit to Florence, he saw and conversed with the great Galileo." Charles Symmons, The Life of John Milton (1822; reprint, London: G. and W. B. Whitaker, 1970), pp. 82–83.

11. John Milton, "Il Ponseroso," in *Paradise Lost and Selected Poetry and Prose,* ed. Northrop Frye (New York: Holt, Rinehart and Winston, 1951), p. 322, lines 31–39. The idea of Galileo's daughter and "Il Ponseroso" is from Fanny Byse, *Milton on the Continent: A Key to L'Allegro and Il Ponseroso* (1903; reprint, Folcroft, Pa.: The Folcroft Press, 1969), pp. 39–45.

12. Milton, *Paradise Lost,* in *Paradise Lost and Selected Poetry and Prose,* p. 115, bk. 5 lines 257–66.

13. Ibid., pp. 178–79, bk. 8 lines 15–38.

14. William Wordsworth, "Star Gazers," in *Wordsworth: Selected Poems*, ed. Herschel M. Margoliouth (London: Collins, 1959), p. 469.

15. Alfred, Lord Tennyson, "In Memoriam," in *Victorian Poetry and Poetics*, ed. Walter E. Houghton and G. Robert Stange (Boston: Houghton Mifflin Co., 1968), p. 76, canto 106, lines 1–8.

Chapter Four

1. Tennyson, "The Princess: Now Sleeps the Crimson Petal," in *Victorian Poetry and Poetics*, p. 44, lines 7–10.

2. Milton Millhauser, *Fire and Ice: The Influence of Science on Tennyson's Poetry* (Lincoln: The Tennyson Society, Tennyson Research Center, 1971), p. 19.

3. *Victorian Poetry and Poetics*, p. 6.

4. Tennyson, *In Memoriam*, p. 78, canto 114, lines 1–4.

5. Ibid., p. 80, canto 120, lines 5–8.

6. J. W. Pearce, *In Memoriam* (New York: Macmillan, 1912), p. xl.

7. K. W. Gransden, "The Poet and the Age," in *Tennyson: In Memoriam*, (London: Edward Arnold Publishers, 1964), pp. 11, 19.

8. *In Memoriam*, p. 46, canto 5, lines 1–4.

9. Ibid., p. 54, canto 34, lines 5–8.

10. Ibid., pp. 69–70, canto 34, lines 43–52.

11. Ibid., p. 46, canto 3, lines 1–16.

12. Ibid., pp. 59–60, canto 56, lines 9–20.

13. Hopkins, "Nondum," *The Poetical Works of Gerard Manley Hopkins*, pp. 91–92, 289. I thank Dr. Norman MacKenzie for suggesting to me this similarity of thought between Hopkins and Tennyson.

14. Millhauser, *Fire and Ice*, p. 19.

15. *In Memoriam*, p. 82, canto 128, lines 5–6.

16. Ibid., p. 51, canto 21, lines 17–20.

17. Satellite data from J. Kelly Beatty, Brian O'Leary, and Andrew Chaikin, *The New Solar System* (Cambridge, Mass.: Sky Publishing, 1980), p. 220.

18. *In Memoriam*, p. 51, canto 24, lines 1–4.

19. Ibid., p. 51, canto 23, lines 13–16.

20. Ibid., p. 59, canto 54, lines 1–20.

21. Ibid., p. 79, canto 118, lines 1–28.

22. Tennyson, "Maud," in *Victorian Poetry and Poetics*, p. 91, lines 136–43.

23. *In Memoriam*, p. 81, canto 123, lines 1–8.

24. Ecclesiastes 1:7. Although some sources suggest that Solomon himself authored Ecclesiastes, Martin Luther and others date the book to a later author. However, since fragments of the text were found at Qumran, it could not have been dated later than 150 B.C.E. See C. Ryrie, *The Ryrie Study Bible* (Chicago: Moody Press, 1976).

25. *In Memoriam*, p. 80, canto 121, lines 1–4, 9–12, 17–20.

26. Tennyson, "Locksley Hall," in *Victorian Poetry and Poetics*, p. 33, lines 7–10.

27. Matthew Arnold, "A Summer Night," in *Victorian Poetry and Poetics*, p. 446, lines 76–92.

Chapter Five

1. This chapter is adapted from my thesis, *The Starlight Night: Hopkins and Astronomy* (Department of English: Queen's University, 1979); and subsequent paper "Gerard Manley Hopkins and Some Mid-19th Century Comets," *Journal of the Royal Astronomical Society of Canada* 75, no. 3 (1981): 139–50.

2. The weather conditions from that night were from the London *Times*, August 4, 1864. Solar system data is courtesy of Michael Payette, and was checked using *The Sky* Graphical Astronomy Software Version 1.04, (Golden, Colo.: Software Bisque, 1992).

3. Norman MacKenzie, *The Poetical Works of Gerard Manley Hopkins* (Oxford: Clarendon Press, 1990), p. 40.

4. Ibid., pp. 244–46, personal communication.

5. *Cornhill* (October 1862): 550–51.

6. *London Times*, 1 August 1864, 5e.

7. *London Review*, 13 July 1861, 44.

8. *London Times*, 3 July 1861, 5e.

9. Amedée Guillemin, *The World of Comets*, trans. and ed. James Glaisher (London: Sampson Low, 1877), pp. 254–55.

10. Judges 6:36–37.

11. Guillemin, *The World of Comets*, p. 314.

12. MacKenzie, *The Poetical Works of Gerard Manley Hopkins*, p. 40.

13. Guillemin, *The World of Comets*, p. 32.

14. J. Russell Hind, *The Comet of 1556; being popular replies to every-day questions, referring to its anticipated reappearance, with some observations on the apprehension of danger from comets* (London: John W. Parker and Son, 1857) p. 1.

15. Ibid., p. 14.

16. *London Times*, 4 July 1861, 12; repeated verbatim in *Illustrated London News*, 6 July 1861, p. 3.

17. Hind, *The Comet of 1556*.

18. Hopkins, *The Journals and Papers of Gerard Manley Hopkins*, (entry for 13 July 1874) ed. Humphrey House and Graham Storey (London: Oxford University Press), p. 249.

19. Hopkins, *Further Letters of Gerard Manley Hopkins Including his Correspondence with Coventry Patmore*, (letter of December 1878), ed. Claude C. Abbott (London: Oxford University Press, 1956), pp. 135–36.

20. Ibid., p. 137.

21. Coventry Patmore, *Poems* (London: George Bell, 1906) p. 255.

22. Hopkins, *Further Letters*, p. 317.

23. Brian Marsden, "The Sungrazing Comet Group," *Astronomical Journal* 72 (1967): 1179.

Chapter Six

1. Henry David Thoreau, *The Heart of Thoreau's Journals,* ed. Odell Shepard (New York: Dover, 1927, 1961), undated entry.

2. Ralph Waldo Emerson, "The American Scholar," in *Selected Essays, Lectures, and Poems,* ed. Robert Richardson (New York: Bantam, 1990), p. 84.

3. Emerson, "Self-Reliance," *Selected Essays,* p. 155.

4. Thoreau, *Journals,* undated, 1840s.

5. Thoreau, *Walden* (New York: Peebles Press International, n.d.), pp. 7–8.

6. Venus data was accessed through *The Sky,* Graphical Astronomy Software.

7. Thoreau, *Walden,* p. 23.

8. Thoreau, *Journals,* 26 July 1840.

9. Ibid., 19 April 1854.

10. Ibid., 13 September 1852.

11. William Graves Hoyt, *Planets X and Pluto* (Tucson: University of Arizona Press, 1980), p. 12.

12. Ibid., p. 15.

13. Thoreau, *Journals,* 13 September 1852.

14. Ibid., 13 February 1852.

15. Ibid., 18 February 1852.

16. Walt Whitman, "When I heard the learn'd astronomer," in *The American Tradition in Literature,* vol. 2, p. 102.

17. Thoreau, *Journals,* 4 October 1859.

18. Ibid., 28 October 1852.

19. Milton, *Paradise Lost,* p. 85, bk 4 line 208.

20. Ibid., p. 86, bk. 4 lines 248–51.

21. Ibid., p. 96, bk. 4 lines 604–609.

22. Tennyson, "Ulysses," in *Victorian Poetry and Poetics,* p. 31.

Chapter Seven

1. W. B. Yeats, "The Second Coming," *Collected Poems of W. B. Yeats* (London: Macmillan, 1977), p. 210.

2. Thoreau, *Walden*, pp. 285, 287.

3. *Othello*, 5.2.107.

4. Tennyson, "Morte d'Arthur," in *Victorian Poetry and Poetics*, p. 28, lines 11–12.

5. Frost, "Kitty Hawk," in *Collected Poems, p.* 445.

6. Eugene M. Shoemaker to Brian G. Marsden, 26 March 1993.

7. John Keats, "On First Looking Into Chapman's Homer," in *The Poems of John Keats*, ed. H. W. Garrad (London: Oxford University Press, 1966), p. 38.

8. T. S. Eliot, "The Hollow Men," in *Collected Poems: 1909–1062* (London: Faber and Faber, 1974), p. 92.

9. Percy Bysshe Shelley, "Epipsychidion," in *Selected Poetry and Prose of Shelley*, ed. Carlos Baker (New York: Modern Library of Random House, 1951), p. 283, lines 368–73.

10. Frost, "Fire and Ice," in *Collected Poems*, p. 204.

11. Coleridge, "Kubla Khan," in *English Romantic Poetry and Prose*, p. 391. Coleridge wrote this poem in 1797 after experiencing a hallucination from two grains of opium taken to control an attack of dysentery.

12. Frost, "The Road Not Taken," in *Collected Poems*, p. 103.

13. Yeats, "Easter 1916," in *Collected Poems*, p. 205. The poem commemmorates the Irish rebellion of Easter Sunday of that year; the British executed the people named in the poem.

I write it out in a verse—
MacDonagh and MacBride
And Connolly and Pearse
Now and in time to be,
Wherever green is worn,

Are changed, changed utterly:
A terrible beauty is born.

Epilogue

1. Sir John Davies, *Nosce Teipsum*, in ed. E. Hershey Sneath, *Philosophy in Poetry: A Study of Sir John Davies's Poem "Nosce Teipsum"* (1903; reprint New York, Greenwood Press, 1969), p. 232.

2. Milton, *Paradise Lost*, p. 180, bk. 8 lines 15–38.

3. Ibid., p. 181.

4. L. Alvarez, W. Alvarez, F. Asaro, and H. Michel, "Extraterrestrial Cause for the Cretaceous-Tertiary Extinction," *Science* 208 (1980): 1095–1107. Walter Alvarez's book *T Rex and the Crater of Doom* (Princeton: Princeton University Press, 1997) is an excellent and readable account of the complex story that led to our understanding of what happened on the last day of the Mesozoic era.

5. Percy Bysshe Shelley, *A Defence of Poetry*, in ed. Carl R. Woodring, *Prose of the Romantic Period* (Boston: Houghton Mifflin, 1961), p. 493.

6. Peter Jedicke, *The Comet Hunter's Call*, unpublished manuscript, 1995. Used by permission.

List of Illustrations

Index